SUPERFLUIDITY AND SUPERCONDUCTIVITY

To Pat and Jean

Superfluidity
and
Superconductivity

David R. Tilley
and
John Tilley

Department of Physics
University of Essex

A HALSTED PRESS BOOK

JOHN WILEY & SONS
New York

Published in the USA by Halsted Press, a Division of
John Wiley & Sons, Inc., New York.

Library of Congress Cataloging in Publication Data

Tilley, David R
    Superfluidity and superconductivity.

    'A Halsted Press book.'
    Bibliography: p.
    1. Superfluidity. 2. Superconductivity.
I. Tilley, John, joint author. II. Title.
QC175.4.T54   530.4   74-7081
ISBN 0-470-86788-4

Printed in Great Britain by
Butler & Tanner Ltd, Frome and London

Preface

For some time it has been recognized that there is a marked similarity between superfluid helium and superconductors. The chief characteristic of both is their ability to sustain particle currents at a constant velocity for long periods of time without any driving force. The currents involve the flow of large numbers of particles, and are the only known examples of the motion of systems of macroscopic size which is not quickly destroyed by dissipative processes. It was first emphasized by London, in his well-known books, that the supercurrents should be regarded as quantum currents, and nowadays this interpretation is supported by the experimental results on quantization of circulation and flux.

To describe the condensed particles in either system, we use a macroscopic wave function. The wave function is a complex quantity: its amplitude squared is proportional to the density of the superfluid, or to the number density of super-electrons: the gradient of its phase is proportional to the velocity of the superfluid, or supercurrent. In this book we have attempted to show how the macroscopic theories of liquid helium and superconductors can be developed in parallel from this same starting point.

We have had in mind particularly the postgraduate student beginning research in either experimental or theoretical low-temperature physics. We believe that anyone doing research in either superconductivity or liquid helium would benefit from knowing the relationships between the two. Already, there are a number of examples where discoveries in the one field have stimulated work in the other. In addition, we hope that parts of the book will prove useful to undergraduates taking courses in low-temperature physics.

We have kept the mathematical content at a low level, in order to make the book more widely accessible. We have not attempted to deal with the microscopic theory of superfluidity or superconductivity in detail, but a short account of the BCS theory can be found in the Appendix; there is as yet no corresponding microscopic theory of liquid helium. We have carried the book as far as a fairly complete introduction to the Ginzburg–Landau theory, which we hope will prove useful.

There are problems at the end of most chapters, which have been chosen mainly to amplify the text. The equations are in SI form, but we have not been pedantic about quoting experimental quantities in SI units. In particular, we give magnetic fields in tesla or gauss, not in $A\,m^{-1}$.

In writing the book, D.R.T. has been very conscious of the debt he owes to his friends and collaborators over the years, particularly to John Baldwin and Terry Clark, with whom he learned most of what he knows about the subject.

Likewise, J.T. is indebted to Christopher Matheson, friend and colleague for many years. He also wishes to acknowledge the hospitality of Professor Bob Dingle and the School of Natural Philosophy, University of St. Andrews, during a period of study leave, when part of the early work on this manuscript was done.

We should like to thank Brian Saunders, who read most of the first draft of the book and made some useful comments which led to improvements. We are grateful to Cynthia Williams for typing the first draft, and especially to Maggie Coffey for her rapid and accurate typing of the final manuscript. Finally, we thank our wives for their support and encouragement.

Contents

Chapter 1

Superfluids and Macroscopic Quantization

1.1 Basic properties of liquid He^4

The two isotopes of helium have the lowest normal boiling points of all known substances, 4·21 K for He^4 and 3·19 K for He^3. When the temperature is reduced further, both He^3 and He^4 remain liquid under the saturated vapour pressure, and would apparently do so right down to absolute zero. To produce the solid phases requires application of a rather high pressure, 25 atmospheres or more (Figs. 1.1 and 1.2).

This reluctance of helium to condense arises from a combination of two factors, the low mass of the atoms and the extremely weak forces between them. The forces are weak because of the simplicity and symmetry of the helium atom with its closed shell of two electrons and the absence of dipole moments except for the small magnetic moment of the He^3 nucleus. The effect of low atomic mass is to ensure a high value of zero-point energy, as may be seen from the following argument.

At a given instant of time, one particular atom in liquid He^4 occupies a certain volume bounded by the atoms immediately surrounding it. Owing to the motion of the atoms, this volume varies, but we can say that, on average, the atom is contained within a sphere of volume equal to the atomic volume V_a, and that the sphere has radius $R \sim V_a^{1/3}$. From the quantum-mechanical uncertainty relation, it can be inferred that a particle inside such a cavity has an uncertainty in its

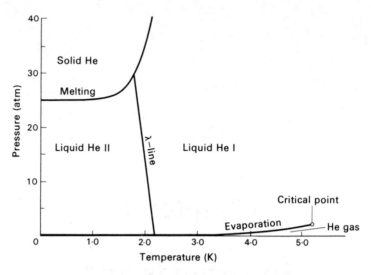

Fig. 1.1   Phase diagram of He^4 (after London 1954).

momentum $\Delta p \sim h/R$, and, consequently, that it possesses kinetic energy of localization, or zero-point energy, $E_0 \sim (\Delta p)^2/2m_4 \sim h^2/2m_4 R^2$, where m_4 is the mass of a He^4 atom. In terms of the atomic volume $E_0 \sim h^2/2m_4 V_a^{2/3}$ and this dependence of E_0 upon V_a is shown schematically in Fig. 1.3. Calculation of the potential energy of the liquid is not easy, and depends upon the choice of model interaction

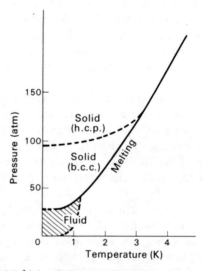

Fig. 1.2   Phase diagram of He^3 (after Grilly and Mills 1959). Hatched area shows region of negative expansion coefficient.

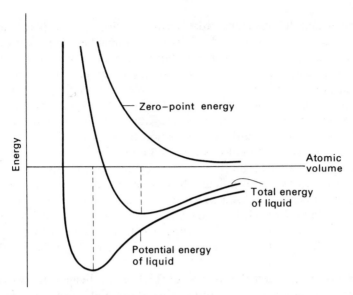

FIG. 1.3 Energy of liquid helium. Total energy is sum of potential energy and zero-point energy.

between two atoms, but it will have the general form of the lowest curve in Fig. 1.3. Because m_4 is small, the zero-point energy is comparable in magnitude to the minimum in the potential-energy curve. The result is that the total energy of the liquid reaches a minimum at a considerably greater atomic volume than the potential-energy minimum. Thus the interatomic forces are strong enough to produce the liquid phase at a low enough temperature, but the high zero-point energy keeps the density of the liquid rather small.

This qualitative argument may be extended to the formation of solid helium. The potential minimum for a lattice will occur at smaller atomic volume than for the liquid (Fig. 1.3), but here the zero-point energy is so large that the lattice is unstable unless a large external pressure is applied. The arguments we have used apply equally to liquid He^3, which has a lower atomic mass and in which the effects of zero-point energy are consequently even greater. Hydrogen is not comparable because the H_2 molecule is much more easily polarized than the single He atom, with the result that the van der Waals force between two H_2 molecules is twelve times stronger than that between two He atoms. In hydrogen, therefore, the binding forces far outweigh the zero-point energy and the solid phase is the stable one at absolute zero. Since all other substances are heavier than hydrogen and have stronger van der Waals interactions, it follows that helium is unique in remaining liquid at indefinitely low temperatures.

Immediately below their respective boiling points, both He^3 and He^4 behave as ordinary liquids with small viscosity. However, at 2·17 K liquid He^4 undergoes a change which is not shared by He^3. This transition is signalled by a specific heat anomaly, whose characteristic shape (Fig. 1.4) has led to the name λ-point being given to the temperature (T_λ) at which it occurs. Furthermore, observation of the liquid at the instant that its temperature is reduced below T_λ reveals a remarkable alteration in its appearance. Liquid helium is maintained at temperatures below 4·2 K by lowering the vapour pressure above the helium bath so that boiling occurs under reduced pressure. Above T_λ bubbles of vapour form within the bulk of the liquid in the customary way and the whole liquid is violently agitated as these rise to the free surface and escape. On the other hand, as soon as the transition point is reached, the liquid becomes quite still and no more bubbles are formed. We infer that T_λ marks the transition between two different forms of liquid He⁴, known conventionally as Helium I above the λ-point and Helium II below it. On the phase diagram (Fig. 1.1) the regions in which the two forms are stable are separated by a broken line, which is not quite vertical, indicating that the transition temperature is lowered when the pressure is increased. The fact that He II is very different from He I, liquid He³ and all other liquids, will become clear as we describe its thermal and flow properties. In §1.3 we shall return to the λ-transition, which occupies a crucial place in the macroscopic theory of liquid He⁴.

Experiments to determine the *viscosity* of He II can be divided into two classes: those designed to measure viscous resistance to flow, and those which detect the viscous drag on a body moving in the liquid. The results shown in Fig. 1.5 are typical of the former; the flow velocity through narrow channels of width varying between 0·1 μm and 4 μm is found to be almost independent of the pressure gradient along the channel. This suggests that the viscosity of He II is virtually zero, a conclusion that is supported by the persistent-current experiments of Reppy and Depatie (1964). In these a torus-shaped vessel was packed with porous material to provide very narrow channels for the liquid. The torus was rotated about its axis of symmetry and then brought to rest, after which the He II continued to flow, showing no reduction in angular velocity over a twelve-hour period, and indicating that He II can flow without dissipation.

On the other hand, experiments using oscillating disks (e.g. Keesom and MacWood, 1938), vibrating wires (Tough, McCormick and Dash, 1963), and rotation viscometers (e.g. Woods and Hollis Hallett, 1963; Fig. 1.6 in this book) demonstrate the existence of a viscous drag, consistent with a viscosity coefficient not much less than that of He⁴ gas. It seems that He II is capable of being both viscous and non-viscous at the same time. This apparent contradiction is the essence of the *two-fluid model*, first suggested by Tisza (1938), in terms of which many of the properties of He II can be explained. According to this model, He II behaves as if it were a 'mixture' of two liquids, one, the normal fluid, possessing an ordinary viscosity, and the other, the superfluid, being capable of frictionless flow past obstacles and through narrow channels. To avoid any misunderstanding, it must be clearly stated at the outset that the two fluids cannot be physically separated; it is not permissible even to regard some atoms as belonging to the normal fluid and the remainder to the superfluid, since all He⁴ atoms are identical. We therefore state the assumptions of the two-fluid model in the following way.

Below T_λ, liquid He⁴ is capable of two different motions at the same instant. Each of these has its own local velocity, respectively $\mathbf{v}_n$ and $\mathbf{v}_s$ for the normal fluid

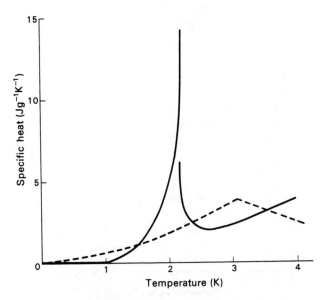

FIG. 1.4 Specific heat of liquid He⁴ (after Atkins 1959). Broken line shows specific heat of ideal Bose–Einstein gas having same density as liquid He⁴.

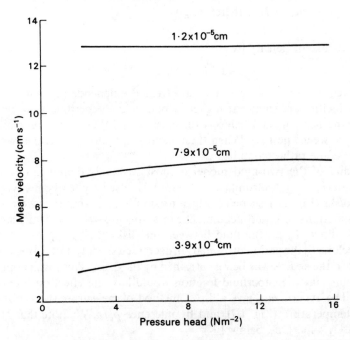

FIG. 1.5 Flow velocity of He II through narrow channels of various widths at 1·2 K. (After Allen and Misener 1939 and Atkins 1952.)

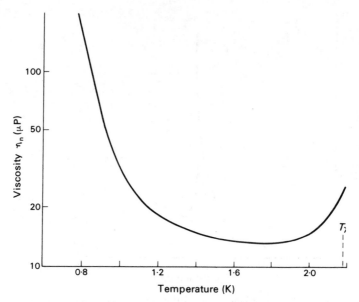

FIG. 1.6   Viscosity of He II measured in a rotation viscometer (Woods and Hollis Hallett 1963).

and the superfluid; likewise each has its own effective mass density, ρ_n and ρ_s. The total density ρ of the He II is therefore given by

$$\rho = \rho_n + \rho_s \qquad (1.1)$$

and the total current density by

$$\mathbf{j} = \rho_n \mathbf{v}_n + \rho_s \mathbf{v}_s. \qquad (1.2)$$

This approach in which the two fluids are treated independently is most useful when the velocities are small. At higher velocities, the superfluid flow becomes dissipative, the normal fluid exhibits turbulence, and there is the possibility of interaction between the two. When these factors are allowed for, the two-fluid equations become rather complicated.

The validity of the two-fluid model is most strikingly demonstrated in the experiment devised by Andronikashvili (1946). He used a pile of equally spaced thin metal disks (Fig. 1.7), suspended by a torsion fibre so that they were able to perform oscillations in liquid helium. The disk spacing was sufficiently small to ensure that above T_λ all the fluid between the disks was dragged with them. However, below T_λ the period of oscillation decreased sharply, indicating that not all the fluid in the spaces was being entrained by the disks. This result confirmed the prediction that the superfluid fraction would have no effect on the torsion pendulum. The experiment gave a direct method of measuring the variation of ρ_n/ρ with temperature (Fig. 1.7), and by inference ρ_s/ρ. We note that He II is almost entirely superfluid below 1 K.

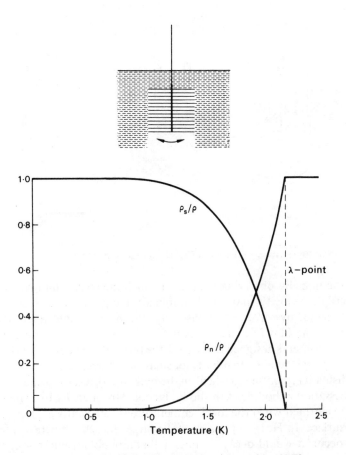

FIG. 1.7   Andronikashvili's experiment (after Atkins 1959).

Another example of the flow properties of liquid He^4 below the λ-point is
provided by the *film* which covers the exposed surface of a body partially immersed in He II. Adsorption on a surface in contact with any liquid or its saturated
vapour is common enough, but He II films are unusually thick. Optical measurements (Jackson and Grimes, 1958) revealed that a typical thickness under saturated vapour is 30 nm or about 100 atomic layers, sufficiently wide to permit
superfluid flow through the film. Owing to the presence of the film on its walls, an
empty beaker lowered into a He II bath begins to fill with bulk liquid, even though
the rim is kept well above the bath surface (Fig. 1.8). Filling continues until the
inner level reaches the level of the bath, at which point it stops. If the beaker is now
raised, it empties itself again, and if it is raised clear of the bath, drops are seen to
fall from the base of the beaker. We conclude that the superfluid fraction flows
through the film whenever there is a height difference between the two bulk liquid
levels. In other words, the film acts like a siphon, the driving force for the
superfluid being provided by the gravitational potential difference between the

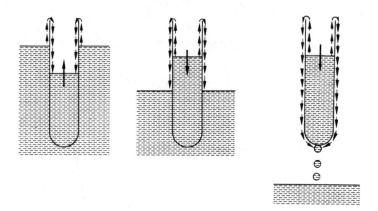

FIG. 1.8   Film flow of He II over the walls of a beaker.

ends of the film. By observing the rate at which the beaker liquid level changes, the superfluid velocity may be determined; a typical value is $20\,\mathrm{cm\,s^{-1}}$. On the other hand, by virtue of its viscosity, the normal fluid fraction is almost stationary in the film. We shall discuss film flow in greater detail in Chapter 2.

Early experiments designed to measure the *thermal conductivity* of He II showed that it is very high, tending to infinity for small heat currents. In fact it is impossible to establish a temperature gradient in the bulk liquid, a result which explains the sudden cessation of boiling as the liquid is cooled through T_λ. In ordinary liquids, a bubble is formed when the local temperature is sufficiently greater than that at the free surface. In He II, supposing that a large enough temperature fluctuation were to occur, it would decay so quickly that a bubble would not appear. Thus evaporation of He II takes place only at the free surface.

A temperature gradient can be set up between two volumes of bulk He II provided that they are connected only by a superleak, that is a channel through which the superfluid can flow, but not the normal fluid. A common form of superleak is a tube packed tightly with fine powder: the spaces between the particles form winding channels of varying width (typically $\sim 100\,\mathrm{nm}$) which allow the superfluid to pass and clamp the normal fluid. If heat is supplied to one side of the superleak, a pressure head is set up as well as a temperature difference (Fig. 1.9). This happens because the superfluid fraction flows from the low-temperature side to the high-temperature side of the superleak. Since ρ_s/ρ increases with decreasing temperature, we infer that the superfluid moves to the region of higher temperature in order to reduce the temperature gradient.

A dramatic demonstration of this effect is furnished by the so-called *helium fountain*, first seen by Allen and Jones (1938) (Fig. 1.10). The superleak in this case is a wide tube containing emery powder. One end is open to the He II bath, while the other is joined to a vertical capillary. When the emery powder is heated, the superfluid flows into the superleak with such speed that He II is forced out of the

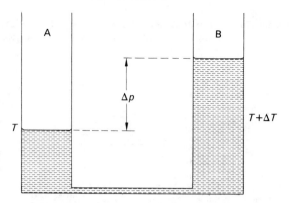

FIG. 1.9 Two vessels connected by a superleak. A temperature difference between the two is accompanied by a pressure head.

capillary tube in a jet. The heat provided by a small hand torch is sufficient to produce a fountain rising to heights of 30 or 40 cm.

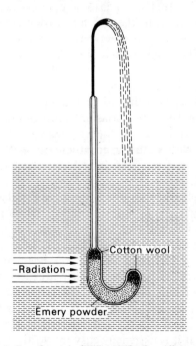

FIG. 1.10 The helium fountain (Wilks 1967, after Allen and Jones 1938).

These manifestations of the *thermomechanical effect* show clearly that heat transfer and mass transfer in He II are inseparable. The steady supply of heat to the bulk liquid, achieved for example by passing direct current through a resistor, and its removal elsewhere into a constant-temperature reservoir causes internal convection (Fig. 1.11). Normal fluid flows from the source to the sink of heat, whilst superfluid flows in the opposite direction, under the constraint that the total density remains constant everywhere. Thus heat is not transferred in He II by the ordinary processes of conduction and simple convection of the whole fluid. Only the normal fluid fraction carries heat; superfluid flow by itself cannot transport heat.

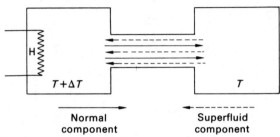

FIG. 1.11   Internal convection in He II. Heat is supplied by heater H and temperatures are held constant.

When the heat supply to He II is made to vary periodically, by passing alternating current through the resistor, the two fluids oscillate in antiphase with one another. Once more, this has no effect on the total density ρ which remains uniform throughout. The result is that the local value of the ratio ρ_s/ρ, and consequently the local temperature, undergo oscillations. In this way He II is able to propagate temperature waves, which are given the name *second sound* to distinguish them from *first sound*, the ordinary longitudinal pressure waves involving fluctuations in the total density at constant temperature. Provided that the rate of heat supply is not too large, and the frequency not too high, second-sound waves are propagated with virtually no attenuation. We shall discuss second sound in more detail after we have introduced the two-fluid hydrodynamic equations in Chapter 2.

The behaviour of He II when set into rotation can be described in terms of the two-fluid model, but this is a much more complicated situation than the properties we have described so far, and, for instance, it is not possible to ignore the interaction between the two components. In a rotating vessel, the normal fluid behaves in the expected way, undergoing *solid-body rotation*. The superfluid fraction appears to do the same, but in reality it experiences vortex motion. A series of vortex lines threads the fluid in the rotating vessel. Superfluid rotates round each line, the angular momentum associated with each vortex being quantized. The occurrence of vortices in the superfluid is not limited to the case of a rotating vessel; indeed it is extremely easy for vortices to be created in many situations involving superflow. Vortices in liquid helium will be discussed fully in Chapter 4.

To conclude this introductory section, we turn our attention to the entropy of He II. Looking again at the phase diagram for He^4 (Fig. 1.1), we see that the melting curve is steep for $T > T_\lambda$, but that it changes its slope rapidly below T_λ, eventually becoming horizontal as $T \to 0$. The gradient of the melting curve is determined by the appropriate Clausius–Clapeyron equation:

$$\frac{\mathrm{d}p_{\mathrm{m}}}{\mathrm{d}T} = \frac{\Sigma_{\mathrm{liq}} - \Sigma_{\mathrm{sol}}}{V_{\mathrm{liq}} - V_{\mathrm{sol}}} = \frac{\Delta\Sigma_{\mathrm{m}}}{\Delta V_{\mathrm{m}}} \tag{1.3}$$

where Σ is entropy and V is volume, and the subscripts have obvious meanings. Above the λ-point $\Delta\Sigma_{\mathrm{m}}$ is large, but immediately below T_{λ} it decreases quickly to become virtually zero for all temperatures below 1 K. In this range, therefore, the liquid cannot lose entropy by solidifying, and the liquid phase is the stable one when the temperature is very close to zero. We conclude that $\Sigma_{\mathrm{liq}} \to 0$ as $T \to 0$, in agreement with the Third Law of Thermodynamics. In view of the experimental evidence that also $\rho_{\mathrm{s}}/\rho \to 1$ as $T \to 0$, we conclude that at absolute zero He II is entirely superfluid and possesses zero entropy. In consequence, it is logical to assume that at finite temperatures the superfluid fraction carries no entropy. Indeed, this is an alternative way of saying that the superfluid can flow without dissipation, since any dissipative process invariably involves entropy production. Thus the entropy of He II is confined to the normal fluid, as might be expected after discussing internal convection, in which the normal fluid is responsible for the transport of heat.

It is clear that the pure superfluid constitutes the ground state of He II. The He^4 atom has a resultant spin of zero, and is therefore a boson; an assembly of He^4 atoms is governed by Bose–Einstein statistics. As is well known, an ideal boson gas of particles with non-zero rest mass exhibits the phenomenon known as the *Bose–Einstein condensation*. At low temperatures, the particles tend to crowd in to the same quantum state, the lowest single-particle energy level of the system, forming a *condensate*. The condensation begins at a certain critical temperature and is complete at absolute zero. It seems certain that liquid He^4 behaves in a very similar way. The λ-point is the temperature which marks the onset of condensation, and the condensate is associated with the superfluid fraction of He II. We shall discuss the Bose–Einstein condensation in §1.3. Before that, in §1.2, we introduce the basic properties of superconductors, and then later on in the chapter we shall describe how the idea of a condensate can be applied to both superfluid helium and superconductors.

1.2 Basic properties of superconductors

Superconductivity is a phenomenon with many features in common with the superflow of He II. One immediate difference is that there are many metals which become superconducting at a sufficiently low temperature, whereas superflow in a liquid is unique to He II. The simplest property of the superconducting state is that it is one in which an electric current, if it is small enough, can flow without a voltage appearing; this is analogous to the superflow of He II through a thin channel or a surface film. Superconductivity is characterized by a critical temperature T_{c}; the resistance of a superconducting wire drops to zero more or less discontinuously at T_{c}. It is believed that the superconducting state really is a state of zero resistance, and not simply a state of very low resistance. An elegant way of demonstrating that the currents do flow without resistance is to suspend a bar magnet above a concave superconducting dish. The induced supercurrents act to repel the magnet, and it stays suspended indefinitely. Shoenberg's book (1952) contains a photograph of this experiment, which of course resembles the persistent superflow of He II in a torus.

The transition to superconductivity is a virtually perfect second-order phase transition; that is, there is no latent heat, and a sharp finite discontinuity in the specific heat. Figure 1.12 shows the specific heat as a function of temperature for Nb, which is typical. This almost ideal behaviour of the specific heat in supercon-

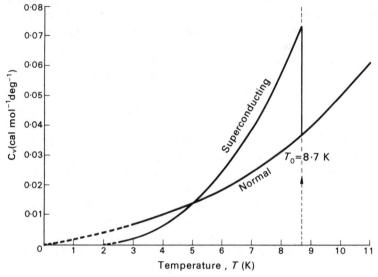

FIG. 1.12   Specific heat of Nb. Normal state values are measured in a magnetic field greater than H_{cb}.
(After Brown *et al.*, 1953.)

ductors contrasts with the λ anomaly in helium (Fig. 1.4). We shall see in Chapter 4 that both the superfluid and superconducting states are characterized by a range of coherence ξ, which however is much shorter in He II than in superconductors; the behaviour of the specific heat is governed by the coherence range, in a way that we shall discuss in §6.9.

Because so many materials undergo a transition to superconductivity, any discussion of the subject is inevitably complicated to some extent by the need to differentiate the behaviour of different classes of superconductors. The most complete tabulation of properties of superconducting materials is that given by Roberts (1971). Table 1.1, taken from that source, shows the elements which become superconducting with a critical temperature above 0·8 K. It can be seen that there is no simple rule to decide which elements become superconductors. However, the following points deserve mention. Firstly, only metals become superconductors. Secondly, all the critical temperatures are under 10 K; actually Table 1.1 does not include the highest critical temperatures, as some metallic compounds have critical temperatures of about 20 K. Thirdly, some metals which are good conductors at room temperature, notably the noble metals, do not become superconductors at all. Fourthly, magnetic metals do not become superconductors.

TABLE 1.1. *Elements which become superconducting above 0·8 K. We shall show critical temperature and, where known, critical field at zero temperature. Different determinations do not always agree, in which case we show an average value. Data from Roberts (1971).*

Element	T_c(K)	H_0 (oersteds)
Al	1·19	100
Ga*	1·09	59
Hg*(α)	4·16	395
In	3·4	288
La*(β)	6·0	1600
Mo	0·92	96
Nb	9·2	1950†
Pa	1·4	
Pb	7·2	800
Re	1·7	193
Sb	2·6	
Sn	3·72	304
Ta	4·4	800
Tc	8	
Th	1·4	145
Tl	2·38	174
V	5·3	1100
Zn	0·86	52

* *More than one phase superconducting. One only shown.*

† *Nb is a type II superconductor. H_{c2} is shown.*

Many metallic compounds become superconductors. In particular, it seems that the cubic β-W(A15) structure is particularly favourable to superconductivity, and the highest critical temperatures of all are found among compounds with this structure.

It is important to be clear about the effects of adding impurities to superconducting elements. Although impurities increase the resistance in the normal state of the metal, they do not prevent the metal becoming a superconductor, and very often the critical temperature is not much altered by alloying. As an example, Fig. 1.13 shows the variation of T_c with impurity content for Sn with In as impurity. There is an initial sharp drop in T_c, then from about 1 percent impurity onwards T_c changes only slowly with impurity content. These characteristics, particularly the initial drop, are those that generally occur; the important exception is that magnetic impurities destroy superconductivity, with T_c dropping to zero at, typically, a concentration of a few percent of the impurity. For many purposes the impurity content of an alloy can be specified by a single parameter, the mean free path l which enters into the low-temperature resistance of the normal state. We shall see later that although alloying does not affect the critical temperature much, it can drastically alter the current-carrying properties of a superconductor.

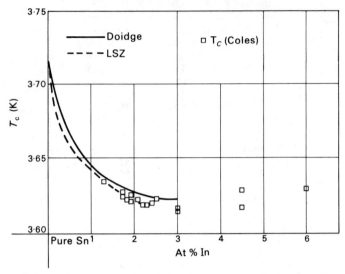

FIG. 1.13   Variation with In concentration of critical temperature of Sn–In alloys. (After Coles 1962.)

Like superflow in He II, a supercurrent is a state of flow without dissipation. An obvious difference is that the current in a superconductor is an electric current, and consequently it generates a magnetic field, as in the 'floating-magnet' experiment. Conversely, an applied magnetic field generates supercurrents, and the most convenient way of looking at supercurrents is to study the behaviour of superconductors in magnetic fields. We now survey the principal magnetic properties.

To begin with, we must make a distinction between the magnetic vectors $\mathbf{H}$ and $\mathbf{B}$ in a superconductor. In dealing with ordinary magnetic materials, one distinguishes between external currents, for example the current in a solenoid around a specimen, which generate $\mathbf{H}$, and *magnetization currents* which affect $\mathbf{B}$ but not $\mathbf{H}$. Examples of the latter are the currents arising from the orbital motions of electrons, and the 'currents' associated with electronic spins. In a superconductor, various conventions are possible, as discussed in the book by Rose-Innes and Rhoderick (1969). We shall always take the view that the supercurrents are magnetization currents, so that $\mathbf{H}$ is unaffected by the presence of the superconductor, whereas $\mathbf{B}$ varies across it according to

$$\operatorname{curl} \mathbf{B} = \mu_0 \mathbf{J}_e,\tag{1.4}$$

where $\mathbf{J}_e$ is the supercurrent density. As usual in electromagnetic theory, the equation $\operatorname{div} \mathbf{B} = 0$ allows us to introduce a vector potential $\mathbf{A}$, with

$$\mathbf{B} = \operatorname{curl} \mathbf{A}.\tag{1.5}$$

The vector potential is particularly important in superconductors; we shall see in later chapters that in a sense the supercurrents respond directly to $\mathbf{A}$.

Various experimental techniques have been used to measure the magnetic behaviour of superconductors. One can measure the integrated voltage produced in a secondary coil when the specimen is removed completely from the applied field; Livingston's (1963) data of Fig. 1.19 were obtained by this means. A refinement of this technique is found in the vibrating sample magnetometer, as used by French and Lowell (1968) for example, in which the specimen is made to oscillate in the applied field. In a torque magnetometer (e.g. Robinson, 1966) the torque between the specimen and a small applied field at right angles to the main field is measured. What all the techniques have in common is that they measure the total flux through the specimen, or equivalently the spatial average of **B**, which we denote by $\langle \mathbf{B} \rangle$. Experimental results are often plotted in terms of the average magnetization **M** rather than $\langle \mathbf{B} \rangle$; **M** is defined by

$$\langle \mathbf{B} \rangle = \mu_0 \mathbf{H} + \mathbf{M}. \tag{1.6}$$

In fact, results are generally plotted as graphs of $-M$ against $\mu_0 H$, since the response is generally diamagnetic.

The simplest type of magnetic behaviour is found in pure metals (except V and Nb), and consists essentially of complete exclusion of magnetic flux B from the specimen for applied fields H less than a critical value H_{cb}. Above H_{cb} there is complete flux penetration, and the normal state is restored. To be precise, B drops to zero over a distance λ, called the penetration depth, at the surface of the specimen, as sketched in Fig. 1.14. λ is typically about 10^{-7} m. For a bulk

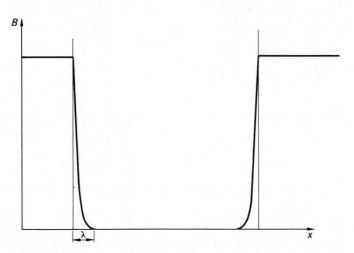

FIG. 1.14   Variation of flux density B across a specimen.

specimen, of dimensions much larger than λ, $\langle B \rangle$ is therefore measured as zero for $H < H_{cb}$. Figure 1.15 is a sketch of the $\langle B \rangle$–H curve, and the corresponding plot of $-M$ against $\mu_0 H$, for this type of behaviour. The magnetization curve of pure Pb is shown as curve A in Fig. 1.19.

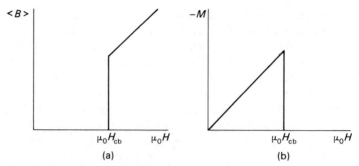

FIG. 1.15   (a) $\langle B \rangle$–$\mu_0 H$ plot, and (b) corresponding magnetization curve, $-M$–$\mu_0 H$, for a type I superconductor.

It can be seen that the shape of the specimen has an important influence upon the magnetic behaviour. Fig. 1.16(a) shows the lines of **B** for a flux-excluding sphere in a uniform applied field. It is clear that the flux density at the equator of the sphere is higher than in the distant applied field; to be precise it is higher by a factor of 3/2. By contrast, if a long ellipsoid is placed in a magnetic field, as in Fig. 1.16(b), the field around the specimen is essentially the same as the distant field.

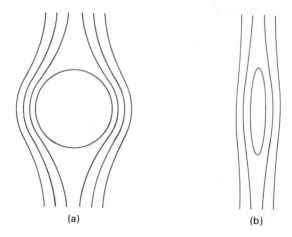

FIG. 1.16   Flux density lines for (a) a diamagnetic sphere, (b) a diamagnetic ellipse.

Now consider what happens when a superconducting sphere is placed in an increasing magnetic field. As long as the applied field is less than $2H_{cb}/3$, the field is less than H_{cb} at all points of the sphere, and consequently the flux is excluded, as in Fig. 1.16(a). When the applied field exceeds $2H_{cb}/3$, some magnetic flux has to penetrate the sphere. In fact the sphere breaks up into alternating regions of superconducting and normal material, with the magnetic field passing through the normal regions. This structure is called the intermediate state, and we stress that its occurrence is a consequence of specimen geometry. Although the study of the intermediate state is of considerable interest, it stands rather apart from the line of development we wish to pursue, and we shall not say any more about the intermediate state. Magnetic measurements are generally carried out on specimens which are long thin cylinders parallel to the applied field; the cylinder is an approximation to the long ellipsoid of Fig. 1.16(b), so that the intermediate state does not occur.

We now return to the magnetic behaviour of a pure superconductor, without geometrical effects, and in particular we wish to discuss the temperature dependence of the critical field H_{cb}. The exclusion of flux from the bulk of a superconductor is called the Meissner effect, after one of its discoverers. What Meissner and Ochsenfeld showed is that the flux will be excluded whether the field is applied before the temperature is lowered through T_c or vice versa, so that the state with flux excluded is a true thermodynamic equilibrium state. This means that one can apply equilibrium thermodynamics to superconductors. Since the applied field $\mathbf{H}$ is an intensive variable, the differential of the internal energy U is

$$dU = T\,d\Sigma + \mathbf{H} \cdot d\mathbf{M}. \tag{1.7}$$

To find the thermodynamic critical field, we equate the Gibbs energies G_n and G_s in the two phases, where

$$G = U - T\Sigma - \mathbf{H} \cdot \mathbf{M} \tag{1.8}$$

with the differential

$$dG = -\Sigma\,dT - \mathbf{M} \cdot d\mathbf{H}. \tag{1.9}$$

For isothermal processes (T constant), we therefore have

$$G(\mathbf{H}) = G(0) - \int_0^{\mathbf{H}} \mathbf{M} \cdot d\mathbf{H}. \tag{1.10}$$

On the assumption that the magnetization is zero in the normal phase, this gives

$$G_n(\mathbf{H}) = G_n(0). \tag{1.11}$$

In the superconducting phase we have $\mathbf{M} = -\mu_0\mathbf{H}$, so that

$$G_s(\mathbf{H}) = G_s(0) + \tfrac{1}{2}\mu_0 H^2. \tag{1.12}$$

At the phase change, $G_s(H_{cb}) = G_n(H_{cb})$, so H_{cb} is given by

$$\tfrac{1}{2}\mu_0 H_{cb}^2 = G_n(0) - G_s(0). \tag{1.13}$$

Since the free energies are equal at T_c, eq. (1.13) implies that H_{cb} tends to zero at T_c, as might be expected. The variation of H_{cb} with temperature is described to within a few percent for all materials by

$$H_{cb} = H_0(1 - t^2) \tag{1.14}$$

with

$$t = T/T_c. \tag{1.15}$$

This temperature dependence is of the form predicted by an early phenomenological theory of superconductivity, the Gorter–Casimir two-fluid model, and it is sketched in Fig. 1.17. The value H_0 of the critical field at zero temperature is typically some hundreds of oersteds; Table 1.1 shows values of H_0 for some superconducting elements.

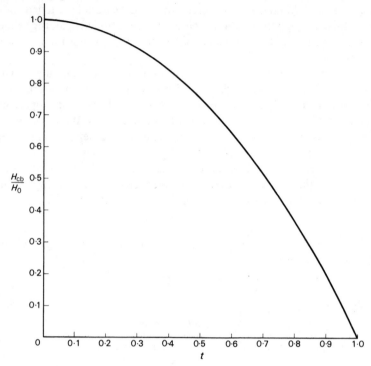

FIG. 1.17   Gorter–Casimir variation of H_{cb} with temperature.

We have already mentioned that the magnetic induction does not drop to zero discontinuously at the surface of a superconductor in the Meissner state; that would imply an infinite current density at the surface. Near a plane surface the induction decays exponentially:

$$B(x) = B(0)\exp\left(-x/\lambda\right), \tag{1.16}$$

where x measures distance into the superconductor from the surface, and λ is the penetration depth. The simplest experimental technique for measuring the penetration depth, originally suggested by Casimir (1940), involves using the specimen as the core of a transformer. The mutual inductance of the primary and secondary coils is governed by the penetration of flux first into the gap between the coils and the specimen, and secondly into the penetration region of the specimen itself. Alternatively, λ can be found from the microwave surface impedance, which we shall discuss in Section 3.3. It is found that λ is temperature dependent, and goes to infinity at T_c. Again, the simple Gorter–Casimir form

$$\lambda = \lambda_0/(1-t^4)^{1/2} \tag{1.17}$$

sketched in Fig. 1.18, fits experimental results to within a few percent, as can be

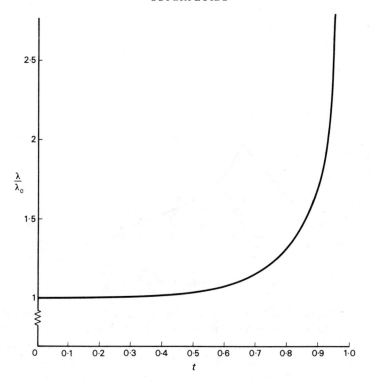

FIG. 1.18   Gorter–Casimir variation of λ with temperature.

seen from Fig. 3.6. Unlike T_c, λ is substantially altered by alloying, increasing as the impurity content increases. An example is shown in Fig. 3.7.

We now turn to the magnetic properties of superconducting alloys. The behaviour is very sensitive to the metallurgical condition of the specimen. The main features of the magnetization curves of well-annealed alloys can be seen from Fig. 1.19, which shows curves taken on a range of Pb–In alloy specimens. We label these magnetization curves as in Fig. 1.20. At low fields the flux is excluded — this is

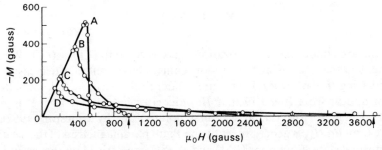

FIG. 1.19   Magnetization curves of annealed polycrystalline Pb and Pb–In alloys taken in ascending field at 4·2 K. A, Pb; B, Pb–2·08 wt % In; C, Pb–8·23 wt % In; D, Pb–20·4 wt % In. (After Livingston, 1963.)

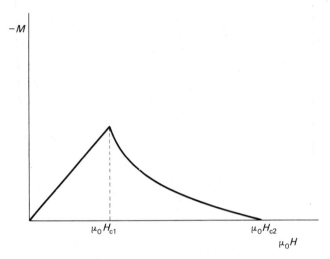

F𝐼G. 1.20   Labelling of type II magnetization curves.

the Meissner region. At a field H_{c1}, the lower critical field, the flux penetrates, but the normal state is not restored completely in the bulk of the specimen until the upper critical field H_{c2} is reached. Furthermore, resistive measurements, as well as theory, show that on surfaces parallel to the applied field the superconducting state persists as a thin ($\sim 10^{-7}$ m) surface sheath up to the surface critical field $H_{c3} = 1{\cdot}69\, H_{c2}$. The region between H_{c1} and H_{c2} is called the mixed state. Superconductors with a magnetization curve containing a mixed-state region, as in Fig. 1.20, are called type II superconductors. As we saw, pure metals (except V and Nb) pass straight from the flux-excluding Meissner state into the normal state, and they are called type I superconductors. Pure V and Nb are type II superconductors, however.

The magnetization curve of a well-annealed type II specimen is almost fully reversible, which implies that the curve passes through equilibrium states. We can therefore apply the thermodynamic eqs. (1.10) and (1.11) to the magnetization curves. Since the Gibbs energies of the two phases are equal at H_{c2}, we find

$$- \int_0^{H_{c2}} \mathbf{M} \cdot \mathbf{dH} = G_n(0) - G_s(0), \qquad (1.18)$$

that is, the area under the magnetization curve is equal to $G_n(0) - G_s(0)$. We note from Fig. 1.19 that increasing the impurity content has the effect of decreasing H_{c1} and increasing H_{c2}, while leaving the area under the curve more or less unaltered. It might appear from Fig. 1.19 that H_{c2} increases indefinitely as the impurity content increases. However, as was first pointed out by Clogston (1962), the ultimate limit on H_{c2} is generally set by the Pauli paramagnetism of the electrons in the normal state. Equations (1.11) and (1.18) were derived on the assumption that the normal state is non-magnetic. In fact, the normal state of most metals is weakly paramagnetic, because in an applied magnetic field it is energetically favourable for the density of electrons with spin aligned along the field to be slightly higher than the density of electrons with spin in the opposite direction. If we allow for this, then eq. (1.11) is replaced by

$$G_n(H) = G_n(0) - \tfrac{1}{2}\chi H^2, \tag{1.19}$$

where χ is the paramagnetic susceptibility. The free energies are equal at a field H_{pl} given by

$$G_n(0) - G_s(0) = \tfrac{1}{2}\chi H_{pl}^2 - \int_0^{H_{pl}} \mathbf{M} \cdot d\mathbf{H}, \tag{1.20}$$

where the final term is the area under the superconducting magnetization curve. This area is always positive, since $\mathbf{M}$ is negative, so we have

$$\tfrac{1}{2}\chi H_{pl}^2 < G_n(0) - G_s(0) \tag{1.21}$$

as a limit on the attainable critical field. Clogston (1962) calculated that in simple materials the numerical value of the limiting field is $\mu_0 H_{pl} = 1{\cdot}84\, T_c$ tesla. The effect of the paramagnetic limiting can be seen on most of the curves for H_{c2} as a function of T shown in Fig. 1.21; the two exceptional cases are substances with a very low paramagnetic susceptibility.

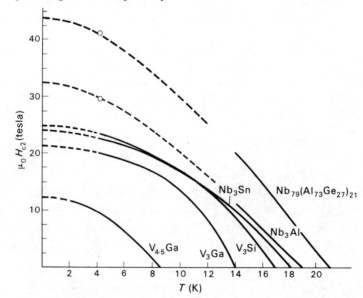

FIG. 1.21 H_{c2} as a function of temperature for various high-temperature, high-critical-field superconducting compounds. The value over 40 tesla is the highest known critical field. (After Foner et al., 1970.)

Type II superconductors which are inhomogeneous, for example because they contain precipitates or a lot of dislocations, have irreversible magnetization curves, as in Fig. 1.22. The curves in increasing and decreasing fields differ appreciably; H_{c1} is no longer clearly defined, but H_{c2} remains well defined. Since the magnetization curve shows hysteresis, it does not pass through equilibrium states, and the area under the curve can be anything at all.

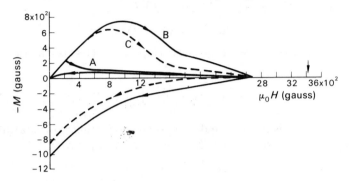

FIG. 1.22   Magnetization curves for a ternary alloy consisting of $78 \cdot 4 \, \text{wt}\%$ Pb, $10 \cdot 0 \, \text{wt}\%$ In, and $11 \cdot 6 \, \text{wt}\%$ Sn. A, As quenched from $260 \, \text{C}$. B, After one day at room temperature. C, After nine days at room temperature. (After Livingston 1964.)

It is nowadays well established that in both reversible and irreversible type II superconductors, the magnetic flux penetrates in the mixed state as a lattice of quantized flux lines, each carrying a flux $h/2e = 2 \cdot 07 \times 10^{-15}$ webers ($= 2 \cdot 07 \times 10^{-7}$ gauss cm^2). The most direct evidence of this comes from the work of Essmann and Träuble (1967), who scattered small cobalt particles on the surface of a superconductor in the mixed state, and made an electron micrograph of the distribution of the particles. As can be seen from Fig. 1.23, the resulting pattern shows the flux line lattice very clearly. In Chapter 4 we shall deal in some detail with the behaviour of quantized flux lines in both superconductors and He II. We

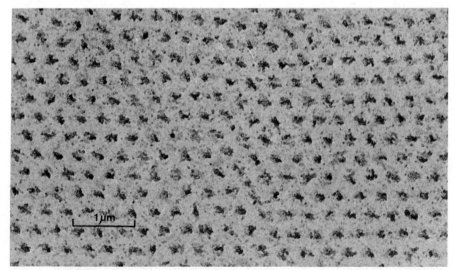

FIG. 1.23   Electron micrograph showing the distribution of small cobalt particles on the surface of a type II superconductor (Pb$_{96}$In$_4$) in the mixed state at $1 \cdot 1 \, \text{K}$. (From Essmann and Träuble 1967.)

shall see there that we can relate the magnetic properties of a superconductor in the mixed state to the behaviour of the flux lines; for example, the irreversibility in magnetization curves like those of Fig. 1.22 is due to pinning of the flux lines by structural defects in the specimen.

As our final topic in this section, we now turn to a qualitative discussion of the microscopic theory of superconductivity. The theory rests securely on the foundations laid in a fundamental paper by Bardeen, Cooper and Schrieffer (1957), and is often referred to as the BCS theory. For He II, by contrast, there is as yet no complete theory. In this book, we aim to treat He II and superconductors in parallel, using macroscopic or phenomenological equations whose primary justification for us is that they give a good description of the experimental situation. Nevertheless, one's confidence in the phenomenological theory for superconductors is increased by some knowledge of the microscopic theory. A full treatment of the BCS theory is beyond our scope, since it requires fairly advanced mathematical techniques. We content ourselves with a review of the main results here, and a fuller description in Appendix I. A good introduction to the detailed theory is to be found in the book by Kuper (1968).

The starting point of the theory is that superconductivity results from an attractive interaction between electrons, mediated by phonons (lattice vibrations). Because of the phonon mechanism, the critical temperature shows an *isotope effect*. In a range of specimens, of Hg say, with different average mass number M, it is found that $T_c \propto M^{-a}$. In simple cases, $a = \frac{1}{2}$. The effect of the interaction on two isolated electrons is to bind them into an entity called a Cooper pair, in which the centre of mass momentum is zero and the electrons have oppositely directed spin. The effect on all the electrons in the metal, at zero temperature, is that they are described by the Bardeen, Cooper and Schrieffer (BCS) ground state, rather than the occupied Fermi sphere of the normal state. The BCS state is made up of Cooper pairs, in the sense that if one electron state of a pair is occupied, then so is the other. It differs from the normal state because some electron states just outside the normal Fermi surface are occupied, and some just inside are unoccupied, as shown schematically in Fig. A1.3. This arrangement requires more kinetic energy than the normal state, of course, but the total energy of the BCS state is lower than that of the normal state, because the binding energy of the Cooper pairs outweighs the increase in kinetic energy.

One of the most important results of the theory concerns the nature of the excited states of a superconductor. A simple excited state of the normal metal consists of a hole of momentum $\mathbf{k}_1$ say, together with an electron of momentum $\mathbf{k}_2$. We may call the hole and the electron *elementary excitations* of the system. The energy required to create the electron–hole pair is $|\varepsilon_{\mathbf{k}_1}| + |\varepsilon_{\mathbf{k}_2}|$, where $\varepsilon_{\mathbf{k}}$ is the energy of the particular electronic state, measured from the Fermi surface. In a superconductor, the nature of the elementary excitations is more complex. An excitation with momentum $\mathbf{k}$ is like the particle in the normal metal if $|\mathbf{k}| \gg k_F$, and like a normal hole if $|\mathbf{k}| \ll k_F$, where k_F is the Fermi momentum. For intermediate values of $\mathbf{k}$ the excitation is part particle, part hole. Furthermore, the energy of an excitation of momentum $\mathbf{k}$ is

$$E_{\mathbf{k}} = (\varepsilon_{\mathbf{k}}^2 + \Delta^2)^{1/2} \qquad (1.22)$$

where, as before, $\varepsilon_{\mathbf{k}}$ is the normal state energy, measured from the Fermi surface. The excitation spectrum is sketched in Fig. 1.24; it can be seen in particular that the minimum energy is $E_{k_F} = \Delta$, so that Δ is the gap in the excitation spectrum. There is some confusion in nomenclature, since some experiments, such as the absorption of light, involve the creation of two quasiparticles. For experiments of this kind, there is a threshold at energy 2Δ, as can be seen for light absorption in Fig. 3.10. *Energy gap* is therefore sometimes used to mean 2Δ. The microscopic theory relates the zero-temperature energy gap $\Delta(0)$ to the critical temperature T_c:

$$\Delta(0) = 1.76 \, k_B T_c . \qquad (1.23)$$

An important way of looking at the excitation spectrum in a superconductor is to consider the density of quasiparticle states $N(E)$, defined as usual so that $N(E)\,dE$ is the number of states between E and $E + dE$. Since there is a gap Δ, N is zero for E between $E_F - \Delta$ and $E_F + \Delta$. The displaced states have to go somewhere, and in fact they are piled up outside the gap region; that is, $N(E)$ is large for E just above $E_F + \Delta$, and just below $E_F - \Delta$. The exact form is shown in Fig. 3.13.

So far we have discussed the BCS state at $T = 0$. We must now deal with the behaviour at finite temperatures. First, the energy gap Δ is temperature dependent, and in fact tends to zero at T_c. The exact form is shown in Fig. 1.25. The physical reason for the temperature dependence of Δ is that the energy gap arises from a cooperative smearing of the Fermi surface which allows the electrons to take advantage of the attractive phonon interaction. However, this cooperative behaviour is undermined by the thermal rounding of the Fermi surface edge, eventually disappearing at T_c with $k_B T_c \sim \Delta(0)$. Secondly, at $T > 0$ there is some thermal excitation of quasiparticles, as sketched in Fig. 3.16. Thus at finite temperatures we have to deal with both the particles in the ground state and a quasiparticle gas. We shall sometimes use the language of a two-fluid model, and refer to these as the superfluid and normal fluid components respectively. It is

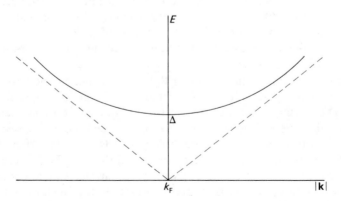

FIG. 1.24   Sketch of BCS excitation spectrum (full line), with normal state spectrum $|\varepsilon_{\mathbf{k}}|$ (broken line). Normal spectrum is $\varepsilon_{\mathbf{k}} = \mathbf{k}^2/2m - k_F^2/2m$, which we approximate by $\varepsilon_k = v_F(|\mathbf{k}| - k_F)$ with $v_F = k_F/m$, since we have $\Delta \ll \varepsilon_F$.

important, however, to realize that a two-fluid model is much less useful for superconductors than for He II. The dramatic two-fluid effects in He II, for instance second sound and the fountain effect, depend essentially upon flow of both the superfluid and the normal fluid components. In a superconductor, however, the normal fluid, or quasiparticle fluid, is scattered by impurities and lattice defects, and consequently is unable to move freely relative to the metallic lattice. This means simply that second sound and the fountain effect do not occur in superconductors.

Once the excitation spectrum is known, the electronic part of the specific heat can be calculated. The energy spectrum in a superconductor, eq. (1.22), is temperature dependent because Δ is temperature dependent. The correct procedure is therefore to calculate the free energy, and differentiate it twice for the specific heat. At low temperatures, $T < 0.5\,T_c$, Δ is essentially independent of temperature, and

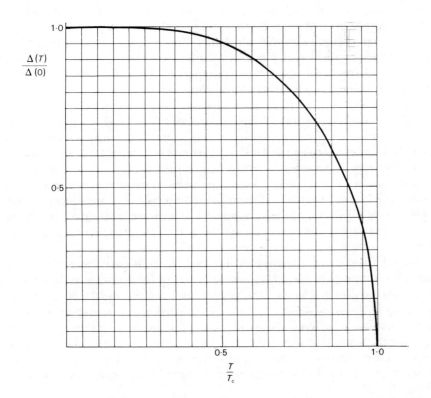

Fig. 1.25 Temperature dependence of the energy gap. (From the table by Mühlschlegel 1959.)

considerably larger than $k_B\,T$. The probability of finding an excitation of energy E_k is then proportional to the Boltzmann factor $\exp\left(-E_k/k_B\,T\right)$, which has a maximum value $\exp\left(-\Delta/k_B\,T\right)$. This factor carries through the calculation of the specific heat, which is given by

$$C_{es} \propto \exp\left(-\alpha\Delta/k_B T\right), \tag{1.24}$$

where α is a numerical factor of order unity. Fig. 1.12 shows satisfactory agreement with this expression. It should be mentioned that the plotting of specific heat results involves subtracting the lattice part of the specific heat, which is not an easy matter. Further details on the specific heat are given by Lynton (1969).

1.3   The Bose–Einstein condensation

In this section we discuss the properties of a model system, the ideal boson gas, and compare them with the properties of liquid He^4. Below a certain critical temperature T_B, the lowest energy level of the ideal gas is occupied by a significant fraction of the particles. We have already noted at the end of § 1.1 the similarity between this behaviour and that of liquid He^4, in which, below the λ-point, a significant part of the liquid becomes superfluid. In order to make a quantitative comparison, we shall find an expression for T_B and calculate the fraction of particles in the ground state at temperatures below T_B.

Before embarking upon the calculation, however, we summarize briefly the reasons why two different kinds of quantum statistics are necessary. All the fundamental particles known at the present time fall into two classes, the distinguishing factor being the intrinsic angular momentum, or spin, of the particle. Spin is quantized and can always be expressed in the form $s\hbar$, where s is the spin quantum number. Those particles which have s equal to an odd half-integer are described by anti-symmetric total wave functions, whereas particles having s equal to an integer or zero require symmetric total wave functions. The symmetry operation referred to here is the exchange of two particles in a system consisting of a set of identical particles. Because they are identical, the exchange of a pair of particles does not alter the physically observable parameters of the system. However, the basic symmetry of the wave functions does determine a fundamental characteristic of the system, namely the number of particles which can occupy the same quantum state.

We take a system consisting of just two identical particles as an example. Suppose that particle 1 is in the state described by the normalized wave function ψ_a, while particle 2 is in ψ_b. The labels a and b stand for the complete set of quantum numbers required to specify a single-particle quantum state. When there is no interaction between the particles, the product $\psi_a(1)\psi_b(2)$ can be used to describe the whole system. On the other hand, because interchanging the particles leaves the properties of the system unaltered, $\psi_a(2)\psi_b(1)$ is an equally valid description. The total wave function is therefore a linear combination of these two products, either

$$\Psi = \frac{1}{\sqrt{2}}\{\psi_a(1)\psi_b(2) + \psi_a(2)\psi_b(1)\} \qquad \text{Symmetric} \quad (1.25)$$

or

$$\Psi = \frac{1}{\sqrt{2}}\{\psi_a(1)\psi_b(2) - \psi_a(2)\psi_b(1)\} \qquad \text{Anti-symmetric} \quad (1.26)$$

Suppose now that both particles are in the same state, that is we put $a = b$. The symmetric total wave function is equal to $\sqrt{2}\psi_a(1)\psi_a(2)$, whilst the anti-symmetric function vanishes. Thus a particle possessing odd half-integer spin is not permitted to enter a state occupied by another identical particle. This is the essence of Pauli's exclusion principle, obeyed by electrons, protons and neutrons amongst others. Although we have discussed only two particles, the argument involving exchange symmetry can be extended to systems containing any number of particles. To take a specific example, the conduction electrons in a normal metal are distributed amongst the available quantum states in such a way that, at any instant, one particular state is either occupied by one particle or empty. Systems with this restriction are governed by Fermi–Dirac statistics, and the constituent particles are called fermions.

In contrast, the fact that the symmetric wave function does not vanish when both particles are in the same state means that particles with integer or zero spin do not obey the exclusion principle. In that case, any number of particles may occupy the same state. Systems of this kind are governed by Bose–Einstein statistics, and their particles are called bosons. The He^4 atom consists of six fermions, but these are bound in such a way that the resultant spin is zero; consequently He^4 is a boson. On the other hand, the He^3 atom comprises five fermions, any combination of which produces a half-integer total spin, so that He^3 is itself a fermion. The two kinds of He atom differ by just one neutron in the nucleus, but this is sufficient to cause them to obey different statistics, which in turn accounts for their disparate behaviour in the liquid phase.

We now return to the model system of an ideal boson gas, consisting of a macroscopic number N of He^4 atoms contained in a box of volume V. In thermal equilibrium at absolute temperature T, the average number of atoms having energy ε_i is given by the Bose–Einstein distribution function

$$\langle n(\varepsilon_i, T)\rangle = \frac{1}{\exp\{(\varepsilon_i - \mu)/k_B T\} - 1}, \tag{1.27}$$

where μ is the chemical potential of the gas. The chemical potential is a thermodynamic parameter which plays a role analogous to that of temperature. Whereas a temperature difference gives rise to flow of energy, a difference in chemical potential results in the flow of particles. We choose the lowest energy level of the gas to have zero energy. In that case we see from eq. (1.27) that μ cannot be positive, otherwise some of the occupation numbers could be negative. In addition, the number of particles in the system is conserved, and so the distribution obeys the sum rule

$$\sum_i \langle n(\varepsilon_i, T)\rangle = N \tag{1.28}$$

at all temperatures. We note the possibility that the number of particles in the lowest energy level $\langle n(0, T)\rangle$ can be of the same order as N, provided that $(-\mu/k_B T)$ is made sufficiently small.

For simplicity the box containing the gas is assumed to be a cube. Under cyclic boundary conditions, the energy levels available to the individual atoms are

$$\varepsilon_{klm} = \frac{\pi^2 \hbar^2}{2m_4 V^{2/3}}(k^2 + l^2 + m^2),$$ (1.29)

where k, l and m are positive integers. The lowest energy level here is ε_{111} which is a positive quantity. To accord with the choice we have already made, that the lowest level should be the zero of energy, we express the energy levels in terms of

$$\varepsilon = \varepsilon_{klm} - \varepsilon_{111}.$$ (1.30)

Furthermore, in a macroscopic system V is so large that the energy levels are very close together, with the result that we treat ε as a continuous variable.

It is now necessary to introduce the *density of states*, $\mathscr{D}(\varepsilon)$, which is the number of energy levels per unit range of energy. After a straightforward calculation using eqs. (1.29) and (1.30) (see problem 1.3), it is found that

$$\mathscr{D}(\varepsilon) = \frac{V}{4\pi^2}\left(\frac{2m_4}{\hbar^2}\right)^{3/2} \varepsilon^{1/2}.$$ (1.31)

The summation over discrete energy levels in eq. (1.27) is now replaced by an integral over all values of ε, that is

$$\sum_i \langle n(\varepsilon_i, T) \rangle \rightarrow \int_0^\infty \mathscr{D}(\varepsilon) \langle n(\varepsilon, T) \rangle \, d\varepsilon.$$ (1.32)

However, this substitution causes a difficulty. The sum includes the particles in the lowest energy level, but the integral does not because $\mathscr{D}(0) = 0$. The integral counts only the particles in the excited levels. The omission of the particles with $\varepsilon = 0$ is serious since, as we have remarked already, $\langle n(0, T) \rangle$ can be of order N. Thus, $\langle n(0, T) \rangle$ must be added on to the integral and, instead of eq. (1.28), we now have

$$N = \langle n(0, T) \rangle + \int_0^\infty \mathscr{D}(\varepsilon) \langle n(\varepsilon, T) \rangle \, d\varepsilon$$

$$= N_0(T) + N'(T).$$ (1.33)

With eqs. (1.27) and (1.31), the integral for $N'(T)$ becomes

$$N'(T) = \frac{V}{4\pi^2}\left(\frac{2m_4}{\hbar^2}\right)^{3/2} \int_0^\infty \frac{\varepsilon^{1/2} \, d\varepsilon}{\exp\{(\varepsilon - \mu)/k_B T\} - 1}.$$ (1.34)

At a given temperature this expression takes its maximum value when $\mu = 0$, so putting $\mu = 0$ gives us an upper bound for $N'(T)$. With $u = \varepsilon/k_B T$, we have therefore

$$N'(T) < N_m'(T) = \frac{V}{4\pi^2}\left(\frac{2m_4 k_B T}{\hbar^2}\right)^{3/2} \int_0^\infty \frac{u^{1/2} \, du}{e^u - 1}.$$ (1.35)

The integral is well-known:

$$\int_0^\infty \frac{u^{1/2} \, du}{e^u - 1} = \Gamma\left(\frac{3}{2}\right)\zeta\left(\frac{3}{2}\right)$$ (1.36)

where
$$\Gamma(s) = \int_0^\infty dt\, t^{s-1} e^{-t} \tag{1.37}$$

and
$$\zeta(p) = \sum_{k=1}^\infty k^{-p}. \tag{1.38}$$

The gamma-function $\Gamma(\tfrac{3}{2})$ has the value $\sqrt{\pi}/2$, and the Riemann zeta-function $\zeta(\tfrac{3}{2}) = 2\cdot612$ (London 1954 (Vol. II), appendix). The maximum number of particles in excited states is therefore

$$N'_m(T) = 2\cdot612\, V\left(\frac{m_4 k_{\mathrm{B}} T}{2\pi\hbar^2}\right)^{3/2}. \tag{1.39}$$

For a system with fixed N and V, at sufficiently high temperatures, $N'_m(T)$ is large enough for all the particles to be accommodated in excited levels. However, as the temperature is reduced, a critical point T_{B} is reached below which $N'_m(T)$ is less than N. In other words, below T_{B} particles start moving into the lowest energy level, and do so in increasing numbers as the temperature is lowered further. The critical temperature T_{B} is defined by putting $N'_m(T_{\mathrm{B}}) = N$, which yields

$$T_{\mathrm{B}} = \frac{2\pi\hbar^2}{m_4 k_{\mathrm{B}}}\left(\frac{N}{2\cdot612\, V}\right)^{2/3}. \tag{1.40}$$

Combination of the last two equations shows that the fraction of particles in excited states is

$$\frac{N'(T)}{N} = \left(\frac{T}{T_{\mathrm{B}}}\right)^{3/2} \qquad (T \leqslant T_{\mathrm{B}}). \tag{1.41}$$

The remainder are in the lowest energy level, and from eq. (1.33) we find

$$\frac{N_0(T)}{N} = 1 - \left(\frac{T}{T_{\mathrm{B}}}\right)^{3/2} \qquad (T \leqslant T_{\mathrm{B}}). \tag{1.42}$$

It follows from eq. (1.42) that $N_0(T_{\mathrm{B}}) = 0$, but this is not exactly true, owing to the approximate nature of the calculation leading to eq. (1.40). In fact, a small fraction of the particles are in the lowest level for $T > T_{\mathrm{B}}$ but it is so small as to be negligible.

Figure 1.26 summarizes the results we have found for the ideal gas. At absolute zero all the particles are in the lowest energy level; above T_{B} almost all the particles are in excited levels. Between absolute zero and T_{B} the particles are divided into two groups, some in the lowest level and some in excited levels. In a system of macroscopic size, the lowest level remains occupied by a macroscopically large number of particles up to a finite temperature. This is the phenomenon known as the Bose–Einstein condensation, the particles in the lowest level comprising what is called the *condensate*. However, this condensation is quite different from that which occurs when, for example, a gas is liquefied. In that case,

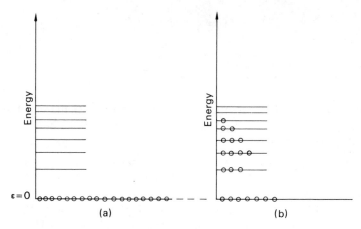

FIG. 1.26   Occupation of energy levels in ideal Bose–Einstein gas (schematic). (a) $T = 0$, all particles in lowest energy level; (b) $0 < T < T_B$, some excited states occupied, but significant fraction of particles remains in lowest level (condensate).

the particles form two phases which are separated by a well-defined boundary in position space. In contrast, the Bose–Einstein condensation can be regarded as a separation in momentum space, but there is no physical boundary between the condensate and the excited particles. Nevertheless, the particles are 'ordered' according to their momenta and from this point of view the Bose–Einstein condensation is an example of an order–disorder transition.

To see how the ideal system relates to helium, we estimate the condensation temperature for gaseous He^4. If in eq. (1.40) we use the particle density N/V appropriate to the saturated vapour at the normal boiling point, we find $T_B = 0.5$ K. Since the boiling point is 4.2 K, the Bose–Einstein condensation is not observed in the gas. However, if we use the density of liquid He^4 instead, we obtain $T_B = 3.1$ K. This is sufficiently close to 2.17 K, the temperature of the λ-transition, to suggest that T_λ marks the onset of Bose–Einstein condensation in liquid He^4. On the other hand, the specific-heat anomalies in the two cases are quite different (Fig. 1.4), the ideal gas showing a cusp singularity in C_V as opposed to the λ-discontinuity of liquid He^4.

It is hardly surprising that the ideal-gas model does not yield an accurate description of the λ-transition, since liquid helium is obviously a system in which the attractive forces between the atoms play an essential part. In the absence of a satisfactory microscopic theory, we make some general assertions about an inter-acting Bose system, whose chief justification lies in the valuable picture of He II that they give.

In comparison with the ideal gas the effects of the interactions are twofold; the number of particles condensed into the lowest energy level is reduced, and the nature of the excited levels of the system is altered. Thus, at absolute zero, not all the particles are in the lowest energy level, but some are raised to levels of slightly higher energy (Fig. 1.27(a)). We say that the condensate is *depleted* by the interactions. The crucial point however is that a finite fraction of the particles still occupies the lowest level, and that this remains true up to a finite condensation temperature. Of course, at all temperatures above absolute zero, the thermally excited levels of the system are occupied to some extent (Fig. 1.27(b)). These no

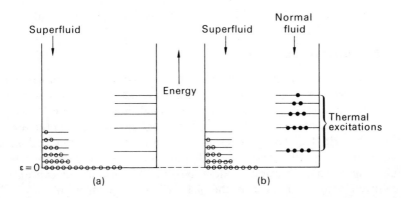

FIG. 1.27 Occupation of energy levels in interacting Bose gas and He II (schematic). (a) $T = 0$, interactions cause some particles to be raised to virtual states with $\varepsilon > 0$; this is known as depletion of condensate. (b) $0 < T < T_B$ (or T_λ), some excited states occupied; owing to interactions, these are not single-particle states. Significant fraction of particles remain in lowest energy level, again depleted owing to interactions.

longer correspond to single-particle states, but to the elementary excitations of the whole system, which, in the first instance, are treated as non-interacting *quasi-particles*.

We can relate this description of an interacting Bose system to the two-fluid model. The experimental evidence is that He II is pure superfluid at absolute zero. Consequently, it appears that the superfluid fraction comprises both the condensate and the particles occupying the depletion levels. At the same time, we identify the normal fluid fraction with the thermal excitations. We shall discuss the condensate in the next section, and the nature of the excitations in He II in § 1.5.

We have already mentioned the similarities between superfluidity and superconductivity, particularly that they both involve the flow of particles without dissipation of energy. At first sight it might seem that the superconducting transition is unrelated to the Bose–Einstein condensation, because in a superconductor the particles involved, electrons, are fermions. However, in deciding the type of quantum statistics that is relevant to a given system, one must take account of the interactions as well as the intrinsic nature of the particles. The simple argument given at the beginning of this section applied strictly to the case with no interaction present. In the case of He^4, there are strong interactions which bind together the constituent particles of the atom, creating a boson, whilst the residual interactions between the atoms are comparatively weak, so that Bose statistics is applicable. Similarly, as we saw in § 1.2, in a superconductor interactions lead to the formation of Cooper pairs of electrons, and these pairs behave sufficiently like bosons to undergo an analogous condensation. The superconducting transition temperature T_c corresponds to the condensation temperature T_B, although the specific heat anomaly at T_c (Fig. 1.12) is different in character from the anomalies in both liquid helium and the ideal Bose gas (Fig. 1.4). Thus below T_c we can regard the Cooper pairs as being condensed into the ground state, while the remaining electrons are in the thermally excited levels of the system.

1.4   The condensate wave function

We have suggested that liquid helium below the λ-point and a superconductor below its transition temperature each possess a condensate, that is, a macroscopically large number of particles occupying a single quantum state. We have also indicated that the condensate possesses the essential properties of the superfluid component of He II, and of the super-electrons in a superconductor. In this section we make the relationship between condensate and supercurrent more definite by introducing a wave function to describe the condensate and expressing the supercurrent velocity in terms of the phase of the wave function.

Apart from supercurrents, the only other currents which flow without dissipation are the orbiting electrons in atomic and molecular systems. Each electron in a stable atom occupies a stationary quantum state, described by an eigenfunction of the appropriate Hamiltonian. In his classic work, London (1954) argued that, by analogy, superfluid currents are also quantum currents, and, furthermore, that some kind of wave function extends throughout a superfluid specimen. To postulate a macroscopic wave function is a very bold step. Atomic and molecular wave functions are confined to regions whose typical size is of the order of a nanometre, whilst, to take an extreme example, the wire in a superconducting solenoid may be many metres long. However, London's suggestion is now supported by a wealth of experimental evidence. We therefore take as our basic premise that superfluids display macroscopic quantization: most of this book will be taken up with exploring the consequences of this postulate. One of our principal aims is to demonstrate the features that He II and superconductors have in common. For that reason, we use the terms superfluid and supercurrent with reference to both.

Up to now we have treated the single quantum state occupied by the particles of the condensate as one of zero energy. Clearly, however, when there is a supercurrent, the quantum state is one with finite energy, namely the kinetic energy of flow. In either case, whether the superfluid is in motion or not, the crucial point is that only a simple wave function is needed to describe the condensate. We postulate that, under steady-state conditions, this wave function can be written in the form

$$\psi(\mathbf{r}) = \psi_0 \exp\{iS(\mathbf{r})\}, \tag{1.43}$$

where the phase $S(\mathbf{r})$ is a real function of position $\mathbf{r}$. In certain situations the amplitude ψ_0 also varies with $\mathbf{r}$, but we shall not deal with these until after we have discussed the interpretation of ψ_0 in Chapters 2 and 3.

By application of the usual operator to $\psi(\mathbf{r})$, we obtain an expression for the condensate momentum $\mathbf{p}$,

$$\hat{\mathbf{p}}\psi = -i\hbar\,\nabla\psi = \mathbf{p}\psi, \tag{1.44}$$

from which it follows that

$$\mathbf{p} = \hbar\,\nabla S. \tag{1.45}$$

The quantity $\mathbf{p}$ defined by (1.45) is the canonical momentum, that is, the momentum conjugate to the position coordinate $\mathbf{r}$ in the sense of Hamiltonian mechanics.

Firstly, we apply (1.45) to He II. In that case it is conventional to interpret the canonical momentum as the momentum of one particle of the superfluid, whence, if the superfluid velocity is $\mathbf{v}_s$, we can write

$$\mathbf{p} = m_4\mathbf{v}_s. \tag{1.46}$$

When this equation is substituted into eq. (1.45), we obtain

$$\mathbf{v}_s = \frac{\hbar}{m_4}\nabla S. \tag{1.47}$$

However, this is not a unique definition of $\mathbf{v}_s$, since it is by no means clear that the atomic mass m_4 is the correct one to use in eqs. (1.46) and (1.47). We shall discuss this point further in §2.3. In the meantime, eq. (1.47) is important because it makes the superfluid velocity proportional to the gradient of the phase of the condensate wave function. Thus, when the superfluid is at rest, the phase has the same value throughout; when the superfluid velocity is finite and constant, the phase varies uniformly in the direction of $\mathbf{v}_s$. We conclude that the phase of the wave function (1.43) is coherent for the whole superfluid. In other words, $S(\mathbf{r})$ is a smoothly varying function of $\mathbf{r}$, not subject to abrupt changes in value over distances of the order of the atomic separation, except near a vortex core. In fact, $S(\mathbf{r})$ is slowly varying even on a macroscopic scale. The effect of phase coherence is to lock the condensate particles together in a state of uniform motion. This may provoke the idea of a rigid structure moving as a whole, but it must be remembered that the 'rigidity' exists in momentum space rather than in position space. Although we shall not discuss the time dependence of the wave function and of $\mathbf{v}_s$ until Chapter 2, the property of phase coherence gives us a qualitative understanding of how a constant superfluid velocity can be maintained for long periods. A sudden change of $\mathbf{v}_s$ would necessarily involve a simultaneous identical alteration in the velocity of each one of a macroscopically large number of particles, an event so unlikely that it can be discounted.

Secondly, we discuss the condensate wave function for a superconductor. We saw in §1.2 that the transition to superconductivity is characterized by the appearance of an energy gap 2Δ in the excitation spectrum. It is an important simplification that Δ is essentially identical to $\psi(\mathbf{r})$ (eq. 1.43). The original formulation of the microscopic theory which we describe in Appendix I and briefly in §1.2, was restricted to situations in which Δ is constant in space. However, Gorkov (1959a, b) was able in an important series of papers to formulate the microscopic theory for situations like the mixed state of a type II superconductor, in which Δ might be expected to vary in space. He showed that the resulting $\Delta(\mathbf{r})$ was complex, and could be regarded as the macroscopic wave function.

In a superconductor the condensate 'particles' are Cooper pairs of mass $2m$ and charge $2e$. In an externally applied magnetic field **B**, with **B** = curl **A**, the canonical momentum of a Cooper pair is given by

$$\mathbf{p} = 2m\mathbf{v_s} + 2e\mathbf{A}, \tag{1.48}$$

where $\mathbf{v_s}$ is the supercurrent velocity (see e.g. Goldstein 1950). If **p** is eliminated between eqs. (1.45) and (1.48), we find

$$\mathbf{v_s} = \frac{\hbar}{2m}\nabla S - \frac{e\mathbf{A}}{m}. \tag{1.49}$$

This equation corresponds to eq. (1.46) for He II; the remarks made earlier about phase coherence and its relation to $\mathbf{v_s}$ apply equally to superconductors.

By taking the curl of eqs. (1.47) and (1.49), we can derive two more equations of significance:

$$\text{curl } \mathbf{v_s} = 0 \quad \text{(He II)} \tag{1.50}$$

$$\text{curl } \mathbf{v_s} + e\mathbf{B}/m = 0 \quad \text{(superconductor).} \tag{1.51}$$

Equation (1.50) is the condition of irrotationality which is satisfied by the superfluid part of He II when undergoing potential flow. We shall use eq. (1.50) in the next chapter, where we describe how the superfluid equation of motion follows from the postulate of the macroscopic wave function. Equation (1.51) is one form of London's equation which accounts for the exclusion of magnetic flux from a superconductor; its application will be discussed in Chapter 3.

Finally, in this section, we note one further consequence of eq. (1.43). It is obvious that $\psi(\mathbf{r})$ must always be single valued. This is achieved when S itself is single valued, or when S is periodic. The latter situation can occur when there is a hole in the superfluid. A trip round a closed contour encircling the hole involves a change in phase of 2π times an integer (or zero). The case when the phase change is finite corresponds to the situation where the superfluid circulates round the hole with a fixed value of angular velocity, implying that condensation into a state of quantized angular momentum is possible as well as the state of uniform linear momentum we have already met. What we have described is vortex motion in the superfluid. In a superconductor, an equivalent argument leads to the quantization of magnetic flux. Both these situations will be treated in Chapter 4.

1.5   Elementary excitations and superfluidity

In the previous two sections we have concentrated on the particles in the ground state of a superfluid system, the condensate. Since both He II and superconductors possess a condensate, and these are the only known systems which display superfluidity, it is logical to argue that the existence of a condensate is a necessary condition for superfluidity to occur. However, it is not a sufficient condition; whether a system having a condensate is also superfluid depends additionally on the nature of the thermally excited states. In this section we shall concentrate on the thermal excitations of He II; the excitations of superconductors have already been introduced in § 1.2, and we shall refer to them briefly at the end of the section.

When discussing the two-fluid model earlier (§ 1.1), we saw that the entire entropy of He II is carried by the normal component. Later, in § 1.3, we identified the normal fluid with the thermal excitations of He II. Clearly, therefore, the thermodynamic parameters are determined by the way in which the excited states are occupied. The energy levels of the ideal He^4 gas are given by the free-particle relation

$$\varepsilon = p^2/2m_4, \tag{1.52}$$

where ε is the energy and p the momentum of a single atom. Of course, this is not applicable to the liquid, in which the interactions give rise to collective motion of the atoms. We must, therefore, consider the normal modes of motion of the liquid as a whole.

The procedure adopted for He II is analogous to that used in dealing with the low-temperature behaviour of a crystalline solid. The normal modes of a crystal are standing waves of sound, and these can be represented by a set of quantized oscillators. An oscillator with angular frequency ω has an energy

$$\varepsilon = n\hbar\omega, \tag{1.53}$$

where n is an integer, and we ignore zero-point energy. As an alternative one may replace the oscillator in its nth excited state by n *phonons* of energy $\hbar\omega$. Thus phonons are quanta of sound; they are the elementary thermal excitations of the crystal. For every different kind of phonon there is a relation between energy and momentum, known as the dispersion relation when expressed algebraically or as the energy spectrum when shown graphically. As a simple example we show in Fig. 1.28 the spectrum for the modes of a 'one-dimensional crystal'. In a real crystal the

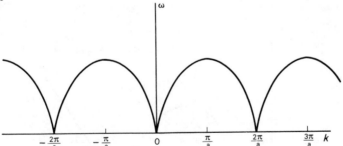

FIG. 1.28   Excitation spectrum of monatomic linear lattice with lattice spacing a.

energy spectrum depends upon the direction of propagation and whether the phonons are associated with transverse or longitudinal vibrations, and consequently the situation can be complex. In principle, knowing the energy spectra, it is possible to calculate the thermal properties of the crystal.

The phonon model of a crystal is an example of a general method of dealing with an interacting system. The original particles and their mutual interactions are replaced by a set of non-interacting, or at least only weakly interacting, *quasi-particles*. At low enough temperatures, the density of quasi-particles is sufficiently small to neglect their interactions. At higher temperatures, they must be taken into account. For instance, the anharmonic behaviour of a crystal lattice can be described in terms of phonon–phonon interactions.

As for liquids, they can propagate longitudinal sound waves, but not transverse ones, so it is to be expected that the thermal excitations of He II should include phonons. The important features of the excitation spectrum of He II were first suggested by Landau (1941, 1947). Initially he proposed that it should consist of two branches (Fig. 1.29), one for phonons and another separated from the first by an energy gap. When he found that this spectrum did not yield the correct thermodynamic behaviour for He II, he replaced it with the continuous spectrum shown in Fig. 1.30. At low energies the curve is a straight line, corresponding to the

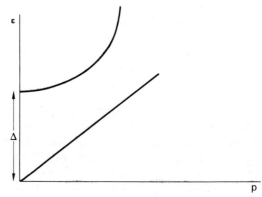

FIG. 1.29   Landau's (1941) initial suggestion for the form of He II excitation spectrum.

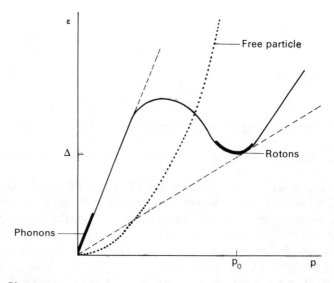

FIG. 1.30   Phonon–roton spectrum suggested by Landau (1947). Broken lines indicate definitions of superfluid critical velocity (eq. 1.62). Dotted line shows free-particle parabola for comparison (eq. 1.52).

phonon dispersion relation

$$\varepsilon = u_1\, p, \tag{1.54}$$

where u_1 is the velocity of sound. At higher energies, the spectrum deviates from a straight line, passing first through a maximum and then a minimum. The excitations with energies near this minimum provide the only major contribution to the thermodynamic parameters besides that of the phonons, and their energy–momentum relation can be expressed in the form

$$\varepsilon = \Delta + \frac{(p - p_0)^2}{2\mu_r} \tag{1.55}$$

where μ_r is an effective mass. We shall see shortly that the existence of the finite energy gap Δ for these excitations, called *rotons*, is crucial for the occurrence of superfluidity in He II. The similarity between Figs. 1.28 and 1.30 suggests that, in a sense, this approach to He II treats it as a solid which has lost some of its rigidity.

The shape of the Landau spectrum has been confirmed by neutron-scattering experiments, particularly those of Yarnell et al. (1959) and Henshaw and Woods (1961). Slow neutrons create and absorb elementary excitations in He II, and by measuring the energy and the deflection of the scattered neutrons, the ε–p curve can be obtained directly. Fig. 1.31 reproduces the results of Henshaw and Woods, and these lead to the following values of the parameters in eqs. (1.54) and (1.55):

$$u_1 = 239\,\mathrm{m\,s}^{-1} \qquad p_0/\hbar = 19 \cdot 2\,\mathrm{nm}^{-1}$$
$$\Delta/k_\mathrm{B} = 8 \cdot 65\,\mathrm{K} \qquad \mu_r = 0 \cdot 16\,m_4.$$

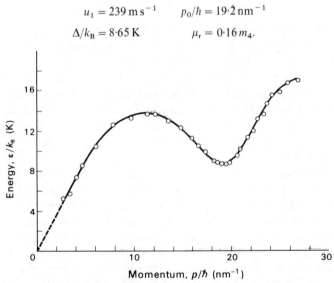

FIG. 1.31 He II excitation spectrum obtained from neutron-scattering experiments (Henshaw and Woods 1961).

We follow the custom of quoting energy in temperature units, and momentum as a wave number. The value for u_1 agrees with that found by direct measurements of the propagation of first sound.

In the case of He II, therefore, the ideal gas of excitations consists of a mixture of phonons and rotons. Their respective contributions to the specific heat may be calculated separately and then added together. For the phonon gas, the Bose–Einstein distribution function is appropriate, but the roton distribution function includes the factor $\exp(\Delta/k_B T)$ which is very large in the He II temperature range and this means that Maxwell–Boltzmann statistics can be used for the rotons. The number densities of phonons and rotons are found to be

$$N_{ph} = 9.60\,\pi\left(\frac{k_B T}{hu_1}\right)^3$$

$$N_r = \frac{2p_0^2(\mu_r\,k_B T)^{1/2}\exp(-\Delta/k_B T)}{(2\pi)^{3/2}\hbar^3}. \tag{1.56}$$

Below 0·6 K the number of rotons excited is negligible and the phonons are the only significant excitations, but above 1 K the rotons play the dominant role from the thermodynamic viewpoint. The phonon and roton contributions to the free energy, entropy and specific heat may also be calculated from the distribution functions and the dispersion relations (1.54 and 1.55). The specific heats of the phonon and roton gases at constant volume are

$$C_{ph} = 0.111\,\pi^4\,k_B N_{ph}/\rho$$

$$C_r = \frac{k_B N_r}{\rho}\left(\frac{\Delta^2}{k_B^2 T^2}+\frac{\Delta}{k_B T}+\frac{3}{4}\right), \tag{1.57}$$

where ρ is the total density of the liquid. For details of these calculations the reader is referred to the books by Khalatnikov (1965) and Donnelly (1967). The latter also contains several useful tables of thermodynamic parameters.

Up to a temperature of 1·2 K the calculated total specific heat is in good agreement with experimental values. Beyond this temperature it is necessary to take account of two additional factors, the growing contributions from the excitations outside the phonon and roton regions, and a slight change in shape of the energy spectrum. In these circumstances, a numerical calculation is required (Bendt et al., 1959).

As one approaches the λ-point the excitation gas model breaks down altogether. Above 1·9 K the roton–roton collision time, for instance, becomes so small that the width of the energy spectrum is comparable with the roton energy (Henshaw and Woods 1961). However, a great deal of information has been obtained from the kinetic theory of the excitations at temperatures well below T_λ, by Landau, Khalatnikov and their co-workers. Analysis of the various collision processes that occur in the mixture of rotons and phonons enabled them to calculate the viscosity coefficient of the normal fluid, and their values are in excellent agreement with, for example, the measurements of Woods and Hollis Hallett (1963) (Fig. 1.6). In addition, Landau and Khalatnikov have calculated the thermal kinetic coefficient and the second viscosity coefficients and thereby have given reasonably accurate theoretical explanations of the absorption of both first and second sound by He II. The theory is summarized by Khalatnikov (1965) and also by Wilks (1967), who discusses its relevance to various experimental results.

The neutron-scattering experiments have been extended to find the shape of the excitation curve at higher values of momentum than in Fig. 1.31. In addition, scattering at high-energy transfer has been interpreted in terms of the creation of more than one excitation (Cowley and Woods, 1971). Besides this, Raman scattering of laser light from He II has indicated that two-roton bound states exist and play a significant role (Greytak et al., 1970).

It is significant that, at temperatures well below the λ-point, the excitation curve proposed by Landau is sufficient to account for the thermal properties of He II. In particular, this fact implies that there are no other excitations in He II with a spectrum lying below the Landau curve; this specifically excludes free-particle motion. We shall now derive the Landau criterion for superfluidity and apply it to the excitation spectrum of He II.

Consider, firstly, He II at 0 K moving through a narrow tube. Superfluid flow is maintained as long as it is slow enough. However, above a certain critical velocity v_{sc}, the atoms move so fast that when they collide with the irregularities in the tube wall, they are removed from the ground state into the excited states. In terms of the quasi-particle model, thermal excitations are created in the liquid with equivalent loss of kinetic energy from the superfluid. On reaching the velocity v_{sc} the flow of He II ceases to be frictionless.

To estimate the value of v_{sc}, we imagine a body of large mass M moving at constant velocity $\mathbf{V}$ through the superfluid which is at rest in the laboratory system of coordinates. The critical velocity v_{sc} is equal to the minimum value of $\mathbf{V}$ at which an excitation can be created. Suppose that the appearance of an excitation with energy $\varepsilon(p)$ and momentum $\mathbf{p}$ causes the velocity of the body to change from $\mathbf{V}$ to $\mathbf{V}_1$. The process must conserve energy

$$\tfrac{1}{2}M\mathbf{V}^2 = \tfrac{1}{2}M\mathbf{V}_1^2 + \varepsilon(p) \tag{1.58}$$

and momentum

$$M\mathbf{V} = M\mathbf{V}_1 + \mathbf{p}. \tag{1.59}$$

If we eliminate $\mathbf{V}_1$ between eqs. (1.58) and (1.59), we obtain

$$\varepsilon(p) - \mathbf{p} \cdot \mathbf{V} + p^2/2M = 0. \tag{1.60}$$

We assume that M is so large that the last term in (1.60) can be neglected. If θ is the angle between $\mathbf{p}$ and $\mathbf{V}$, we then have

$$pV \cos \theta = \varepsilon(p) \tag{1.61}$$

and since $\cos \theta \leqslant 1$, the condition

$$V \geqslant \varepsilon(p)/p$$

must be satisfied for excitations to be created. Thus the critical velocity is given by

$$v_{sc} = \left[\frac{\varepsilon(p)}{p} \right]_{min}. \tag{1.62}$$

Superfluidity can therefore occur if

$$v_{sc} > 0,$$

a condition which is known as the Landau criterion for superfluidity (Landau 1941). Minimum values of $\varepsilon(p)/p$ are found where

$$\frac{d\varepsilon(p)}{dp} = \frac{\varepsilon(p)}{p}. \tag{1.63}$$

There are two solutions of eq. (1.63) on the He II excitation curve. One occurs at the origin, and indeed at all points of the linear part of the spectrum. In this region

$$v_{sc} = \frac{\varepsilon(p)}{p} = u_1 \qquad \text{(phonons)}, \tag{1.64}$$

which indicates that the critical velocity for the creation of phonons is the velocity of first sound, $239\,\text{m s}^{-1}$.

To find the second solution to eq. (1.63), we draw the straight line which passes through the origin and touches the curve close to the roton minimum (Fig. 1.30). From this we obtain

$$v_{sc} \simeq \frac{\Delta}{p_0} = 58\,\text{m s}^{-1} \qquad \text{(rotons)}. \tag{1.65}$$

In Fig. 1.30 we also show the free-particle parabola, corresponding to the dispersion relation (1.52). We see immediately that condition (1.63) can be satisfied on this curve only at the origin, giving

$$v_{sc} = 0 \qquad \text{(free particles)}. \tag{1.66}$$

A critical velocity of zero means that superfluidity is impossible in any system where free-particle motion can take place. Thus it is the energy gap Δ, together with the lack of any other thermal excitations below the Landau curve, which ensures a finite value of superfluid critical velocity in He II.

The limiting velocities of superfluid flow actually observed are much smaller than the value predicted from the excitation curve, sometimes as much as two orders of magnitude less. The reason is that there are other dissipative mechanisms available besides the creation of individual excitations, particularly the motion of vortices (Chapter 4).

In superconductors, the excitation spectrum has an energy gap Δ, as we discussed in §1.2 and sketched in Fig. 1.24; the minimum energy of an excitation occurs at the Fermi momentum $\hbar k_F$. The Landau criterion therefore gives the critical velocity

$$v_{sc} = \Delta(T)/\hbar k_F. \tag{1.67}$$

(The use of Δ for both the superconducting energy gap and the roton energy minimum is so well established in the literature that we think it preferable to retain this double usage, even at a slight risk of confusion.)

We can get a rough idea how eq. (1.67) compares with the experimental values, taking Sn as an example. We treat Sn as a monovalent, free-electron metal (Wilson 1953, p. 249, gives 1·1 free electrons per atom), with $k_F = 1\cdot2 \times 10^{10}\,\text{m}^{-1}$ (Pippard's (1960) *standard metal*). At zero temperature, eq. (1.67) then gives $v_{sc} = 68\,\text{m s}^{-1}$.

We estimate an experimental value of v_{sc} by assuming that a critical current J_{sc} flows in the penetration depth region when the applied field is H_{cb}:

$$J_{sc} = H_{cb}/\lambda. \tag{1.68}$$

Writing $J_{sc} = nev_{sc}$, with the free-electron density n taken as $6 \times 10^{28}\,\mathrm{m^{-3}}$ (Pippard, *loc. cit.*), we find an experimental value at zero temperature of $v_{sc} = 52\,\mathrm{m\,s^{-1}}$. Considering the crudity of the estimate, the two values of v_{sc} are in very satisfactory agreement. As in He II, of course, there can be other dissipation mechanisms besides the breaking up of Cooper pairs; in particular, in the mixed state the motion of the flux lines generally limits the superflow velocity.

Problems

1.1 Solve Schrödinger's equation for a free particle in a spherical cavity, to find the eigenfunctions and corresponding eigenvalues. Hence estimate the zero-point energy of a $\mathrm{He^4}$ atom inside a cavity of volume equal to the atomic volume of liquid $\mathrm{He^4}$ (density $145\,\mathrm{kg\,m^{-3}}$), and compare this with an estimated minimum for the potential-energy curve of $-4 \times 10^{-22}\,\mathrm{J/atom}$ (Fig. 1.3).

1.2 Figures 1.20 and 1.22 show magnetization curves, $-M$ vs. $\mu_0 H$, for reversible and irreversible type II superconductors. Redraw these as curves of $\langle B \rangle$ vs. $\mu_0 H$.

1.3 Derive the density of states function (1.31).

1.4 Estimate the value of the chemical potential of 1 mg of $\mathrm{He^4}$, considered as an ideal gas, at 0·1 K. Compare this with the difference between the lowest and first excited levels, when the gas is contained in a box of macroscopic size.

1.5 Show that a 'two-dimensional' ideal Bose gas does not undergo a Bose–Einstein condensation. (*Hint*: find the density of states for a two-dimensional area.)

1.6 Derive the expressions for the phonon and roton densities (1.56), beginning from the dispersion relations (1.54) and (1.55) and the appropriate distribution function in each case.

References

ALLEN, J. F. and JONES, H., 1938, *Nature, Lond.*, **141**, 243.

ALLEN, J. F. and MISENER, A. D., 1939, *Proc. Roy. Soc.* **A172**, 467.

ANDRONIKASHVILI, E. L., 1946, *Zh. eksp. theor. Fiz.*, **16**, 780.

ATKINS, K. R., 1952, *Adv. Phys.* **1**, 169.

ATKINS, K. R., 1959, *Liquid Helium* (Cambridge University Press, London).

BARDEEN, J., COOPER, L. N. and SCHRIEFFER, J. R., 1957, *Phys. Rev.* **108**, 1175.

BENDT, P. J., COWAN, R. D. and YARNELL, J. L., 1959, *Phys. Rev.* **113**, 1386.

BROWN, A., ZEMANSKY, M. W. and BOORSE, H. A., 1953, *Phys. Rev.* **92**, 52.

CASIMIR, H. B. G., 1940, *Physica* **7**, 887.

CLOGSTON, A. M., 1962, *Phys. Rev. Lett.* **9**, 266.

COLES, B. R., 1962, *IBM J. Res. Dev.* **6**, 68.

COWLEY, R. A., and WOODS, A. D. B., 1971, *Can. J. Phys.* **49**, 177.

DONNELLY, R. J., 1967, *Experimental Superfluidity* (University of Chicago Press, Chicago and London).

ESSMANN, U. and TRÄUBLE, H., 1967, *Phys. Lett.* **24A**, 526.

FONER, S., McNIFF, E. J., Jr., MATTHIAS, B. T., GEBALLE, T. H., WILLENS, R. H. and CORENZWIT, E., 1970, *Phys. Lett.* **31A**, 349.

FRENCH, R. A. and LOWELL, J., 1968, *Phys. Rev.* **173**, 504.

GOLDSTEIN, H., 1950, *Classical Mechanics* (Addison-Wesley, Reading, Mass.).

GORKOV, L. P., 1959a, *Zh. eksp. theor. Fiz.* **36**, 1918 (*Soviet Physics JETP* **9**, 1364).

GORKOV, L. P., 1959b, *Zh. eksp. theor. Fiz.* **37**, 1407 (*Soviet Physics JETP* **10**, 998).

GREYTAK, T. J., WOERNER, R., YAN, J. and BENJAMIN, R., 1970, *Phys. Rev. Lett.* **25**, 1547.

GRILLY, E. R. and MILLS, R. L., 1959, *Ann. Phys. (N.Y.)*, **8**, 1.

HENSHAW, D. G. and WOODS, A. D. B., 1961, *Phys. Rev.* **121**, 1266.

JACKSON, L. C. and GRIMES, L. G., 1958, *Adv. Phys.* **7**, 435.

KEESOM, W. H. and MacWOOD, G. E., 1938, *Physica* **5**, 737.

KHALATNIKOV, I. M., 1965, *Introduction to the Theory of Superfluidity*, tr. by P. C. Hohenberg (W. A. Benjamin, New York).

KUPER, C. G., 1968, *Theory of Superconductivity* (Clarendon Press, Oxford).

LANDAU, L. D., 1941, *J. Phys. Moscow* **5**, 71. (Reprinted in KHALATNIKOV, I. M., 1965, *Introduction to the Theory of Superfluidity*, p. 185, W. A. Benjamin, New York.)

LANDAU, L. D., 1947, *J. Phys. Moscow* **11**, 91. (Reprinted in KHALATNIKOV, I. M., 1965, *Introduction to the Theory of Superfluidity*, p. 205, W. A. Benjamin, New York.)

LIVINGSTON, J. D., 1963, *Phys. Rev.* **129**, 1943.

LIVINGSTON, J. D., 1964, *Rev. Mod. Phys.* **36**, 54.

LONDON, F., 1954, *Superfluids*, Vols. I & II, (reprinted Dover 1964, New York).

LYNTON, E. A., 1969, *Superconductivity*, 3rd. edn. (Methuen, London).

MÜHLSCHLEGEL, B., 1959, *Zeitschrift für Physik* **155**, 313.

PIPPARD, A. B., 1960, *Repts. Prog. Phys.* **23**, 176.

REPPY, J. D. and DEPATIE, D., 1964, *Phys. Rev. Lett.* **12**, 187.

ROBERTS, B. W., 1971, *Handbook of Chemical Physics*, 52nd. edition, ed. R. C. Weast (Chemical Rubber Co., Cleveland, Ohio), p. E73.

ROBINSON, G., 1966, *Proc. Phys. Soc.* **89**, 633.

ROSE-INNES, A. C. and RHODERICK, E. H., 1969, *Introduction to Superconductivity* (Pergamon Press, Oxford).

SHOENBERG, D., 1952, *Superconductivity*, 2nd. edn. (Cambridge University Press, London).

TISZA, L., 1938, *Nature, Lond.*, **141**, 913.

TOUGH, J. T., McCORMICK, W. D. and DASH, J. G., 1963, *Phys. Rev.* **132**, 2373.

WILKS, J., 1967, *The Properties of Liquid and Solid Helium* (Clarendon Press, Oxford).

WILSON, A., 1953, *Theory of Metals*, 2nd. edn. (Cambridge University Press, London).

WOODS, A. D. B. and HOLLIS HALLETT, A. C., 1963, *Can. J. Phys.* **41**, 596.

YARNELL, J. L., ARNOLD, G. P., BENDT, P. J. and KERR, E. C., 1959, *Phys. Rev.* **113**, 1379.

Chapter 2

Helium II:
the Two-fluid Model

2.1 Introduction

We have already indicated in §1.1 that many properties of He II can be understood in terms of the two-fluid model. The basic two-fluid equations were originally put forward by Landau (1941); he treated superfluid flow by expressing the macroscopic hydrodynamical variables in terms of quantum-mechanical operators, and then showing that the equations of motion for these operators were equivalent to the continuity equation and Euler's equation. A derivation of the two-fluid equations starting from a microscopic theory of He II has not yet been achieved. However, the properties of a model system of weakly interacting bosons near absolute zero has been studied extensively. For that system it proves possible to define a *condensate wave function*, and to show that its equation of motion also implies Euler's equation. The procedure we adopt in §§ 2.2 and 2.3 is to assume that we can define a similar wave function for the superfluid part of He II, and treat it in a parallel way.

A theoretical treatment of the weakly interacting Bose gas, involving as it does the use of second-quantization language, is beyond the scope of this book. The interested reader is referred to the relevant articles in the book by Pines (1961), the review article of Martin (1966) and the references therein.

In §2.4 we shall deal with the normal component of He II in the way first put forward by Landau, and in §2.5 we discuss briefly the two-fluid model at higher flow velocities. The remaining sections of the chapter are devoted to an account of experimental work which has largely confirmed the validity of the two-fluid theory.

The paper by Landau (1941) is remarkable for the magnitude of its contribution to our understanding of He II. However, it is interesting to record that, at the beginning of the paper, Landau rejected in no uncertain terms the idea that the properties of an ideal Bose gas below its condensation temperature might have anything to do with the behaviour of He II.

2.2 Basic superfluidity

We begin by considering the condensate. As in § 1.4 we postulate that its behaviour is governed by a single wave function of coherent phase. In the earlier discussion we used a time-independent wave function with constant amplitude. Now we treat the general case, in which both amplitude and phase can vary in space and time,

$$\psi(\mathbf{r}, t) = \psi_0(\mathbf{r}, t) \exp\left[iS(\mathbf{r}, t)\right]. \tag{2.1}$$

We make the assumption that $\psi(\mathbf{r}, t)$ is a solution of a Schrödinger equation

$$i\hbar \frac{\partial \psi}{\partial t} = -\frac{\hbar^2}{2M} \nabla^2 \psi + \bar{V}(\mathbf{r})\psi. \tag{2.2}$$

in which M is a mass to be determined, and $\bar{V}(\mathbf{r})$ is an averaged potential energy. The discussion that follows makes the use of this 'smoothed' potential plausible.

In the elementary problems of quantum mechanics which involve only a single particle, such as the harmonic oscillator and the hydrogen atom, the quantity $\psi^*(\mathbf{r}, t)\psi(\mathbf{r}, t)$ is interpreted as the relative probability of finding the particle at $(\mathbf{r}, t)$. However, when ψ is a macroscopic wave function, $\psi^*\psi$ has a different meaning, as we shall see. In § 1.3 we saw that all the particles of a non-interacting Bose gas would be condensed into the lowest energy level at absolute zero, but that in the presence of interactions the condensate would be depleted. The evidence from theory and experiment, though somewhat meagre, is that only about one-tenth of the particles of He II form the condensate at lowest temperatures (Cowley and Woods 1971). On the other hand, Andronikashvili's experiment (Fig. 1.7) indicated that the relative superfluid density ρ_s/ρ is very close to unity for all temperatures below 1 K. As explained earlier (Fig. 1.27), we assume that the superfluid fraction comprises both the condensate and the depletion. The picture we have is that the particles forming the depletion, although removed from the lowest energy level by the interactions, still follow the motion of the condensate. Consequently, we choose the normalization of $\psi(\mathbf{r}, t)$ so that $\psi^*\psi$ is equal to the average number of superfluid atoms in unit volume, giving

$$\psi^*(\mathbf{r}, t)\psi(\mathbf{r}, t) = \psi_0^2(\mathbf{r}, t) = \rho_s/m_4. \tag{2.3}$$

In the low-temperature limit, $\rho_s \to \rho$, the total density of the liquid; at finite temperatures below T_λ, $\rho_s/\rho < 1$, and at T_λ and above, $\rho_s/\rho = 0$.

Note that eqs. (2.1) and (2.3) imply that both the density ρ_s and the phase S can be determined simultaneously. This is an important step and it is not trivial. In general it is not possible to specify the density and the phase simultaneously in a quantum-mechanical system. For example, in the simple harmonic oscillator a state with N quanta present can have any phase S, and if we specify N precisely we cannot specify S. Conversely, if we specify S, we cannot specify N. Careful examination (e.g., Troup 1967) shows that for large N we can fix both N and S as long as each has an uncertainty given by

$$\delta N \, \delta S \sim 1, \qquad (2.4)$$

so that S can have an uncertainty $\delta S \sim N^{-1/2}$, and N the relative uncertainty $\delta N/N \sim N^{-1/2}$. By analogy, we see that eqs. (2.1) and (2.3) presuppose that the total number of particles comprising the superfluid is large, which of course is the case.

Except in the neighbourhood of a solid wall or a vortex core the superfluid density is a slowly varying function of position, and volume elements in which it is uniform are of macroscopic size. This permits a 'thermodynamic' approach to the description of superfluid motion, for example the use of a slowly varying potential $\bar{V}(\mathbf{r})$ in eq. (2.2). The value of $\bar{V}(\mathbf{r})$ is the potential energy possessed by each one of the macroscopically large number of particles contained in the volume element at $\mathbf{r}$, and $\bar{V}(\mathbf{r})$ is therefore a thermodynamic parameter. We now show that it can be identified with the chemical potential.

A reversible change in the total energy U_0 of a fluid at rest is governed by the identity

$$dU_0 = T \, d\Sigma + \mu \, dN - p \, dV, \qquad (2.5)$$

where the state of the system is determined by the independent variables entropy Σ, number of particles N and volume V. The other quantities in the equation, which is the First Law of Thermodynamics, are absolute temperature T, chemical potential μ and pressure p. We notice that the chemical potential can be expressed as

$$\mu = \left(\frac{\partial U_0}{\partial N} \right)_{\Sigma, V} \qquad (2.6)$$

or, in words, the chemical potential is the energy gained by the system when one particle is added to it at constant volume and entropy. To find the total energy, we must add the kinetic energy of the superfluid $U_{s,k}$ to the energy of the whole fluid at rest U_0,

$$U = U_{s,k} + U_0. \qquad (2.7)$$

Suppose that a volume element containing ΔN superfluid atoms moves from point A to point B in the liquid. Superfluid flow involves no dissipation, so the total energy does not change, nor is there any change of entropy. We can treat the volume element as a sub-system in thermodynamic equilibrium with the rest of the fluid, and apply eqs. (2.5) and (2.7) to it, giving

$$\Delta U = 0 = \Delta U_{s,k} + \left[\left(\frac{\partial U_0}{\partial N} \right)_{\Sigma, V}^{B} - \left(\frac{\partial U_0}{\partial N} \right)_{\Sigma, V}^{A} \right] \Delta N \qquad (2.8)$$

which with eq. (2.6) may be written as

$$\frac{\Delta U_{s,k}}{\Delta N} = -(\mu_B - \mu_A). \qquad (2.9)$$

Hence the chemical potential of the whole fluid at rest acts as the potential energy per particle of the superfluid. When superfluid moves from a region of high μ to one of low μ it is accelerated. In eq. (2.2) we can therefore replace $\bar{V}(\mathbf{r})$ by μ, and the mass M by the mass of a single particle m_4,

$$i\hbar \frac{\partial \psi}{\partial t} = -\frac{\hbar^2}{2m_4} \nabla^2 \psi + \mu \psi. \tag{2.10}$$

Substitution of the wave function (2.1) into eq. (2.10) yields an equation with real and imaginary parts. The imaginary part gives the equation of continuity (see § 2.3 and problem 2.1), whilst the real part is

$$\hbar \frac{\partial S}{\partial t} = \frac{\hbar^2}{2m_4} \frac{\nabla^2 \psi_0}{\psi_0} - \frac{\hbar^2}{2m_4} (\nabla S)^2 - \mu. \tag{2.11}$$

If we use eq. (1.47) for ∇S and eq. (2.3) for ψ_0, we find for the rate of change of the phase

$$\hbar \frac{\partial S}{\partial t} = (\mu + \tfrac{1}{2}m_4 v_s^2) + \frac{\hbar^2}{2m_4} \frac{\nabla^2(\sqrt{\rho_s})}{\sqrt{\rho_s}}. \tag{2.12}$$

The last term is important only when the superfluid density varies rapidly with position, for instance near a wall or inside a vortex. While we are dealing with the bulk fluid we can omit the term.

Equation (2.12) is basic to our understanding of superfluidity. The way in which we have obtained it is not at all rigorous, but the argument presented here has the advantage of illustrating that, although superfluidity is a quantum phenomenon, nevertheless a great deal of progress can be made by describing it in essentially thermodynamic terms.

Anderson (1966) has pointed out that eq. (2.12) can be derived in another way. As can be seen from eq. (2.4), N and $\hbar S$ may be treated as conjugate variables, like the position and momentum of a particle. In the limit of large N, the equations of motion for $\hbar S$ and N take the classical form

$$\hbar \frac{\partial S}{\partial t} = -\frac{\partial \mathcal{H}}{\partial N} \tag{2.13}$$

and

$$\hbar \frac{\partial N}{\partial t} = \frac{\partial \mathcal{H}}{\partial S}, \tag{2.14}$$

where $\mathcal{H}$ is the Hamiltonian. In the case of helium, for $\mathcal{H}$ we can use the total energy U, which includes the superfluid kinetic energy (eq. 2.7). Then, taking mean values, eq. (2.13) becomes

$$\hbar \frac{\partial S}{\partial t} = -\frac{\partial U}{\partial N} = -(\mu + \tfrac{1}{2}m_4 v_s^2) \tag{2.15}$$

which is eq. (2.12) once more, but without its last term.

The importance of eq. (2.15) will emerge as the book proceeds. In § 2.3 we shall show that its gradient is Euler's equation in disguise, that is the equation which governs the hydrodynamics of an ideal fluid. In the case when $v_s = 0$, eq. (2.15) reduces to

$$h\frac{\partial S}{\partial t} = -\mu.$$ (2.16)

In Chapter 4, this relation will be employed to introduce the idea of phase slip, and in Chapter 5 to discuss Josephson effects. Although the treatment in this section has been in the context of helium, eqs. (2.13), (2.14) and (2.16) are, of course, applicable to superconductors. Indeed eq. (2.14) is particularly useful for the description of Josephson effects in superconductors.

2.3 The equation for superfluid flow

First, we shall show how the continuity equation for the superfluid can be obtained from eq. (2.10) and its complex conjugate. The rate of change of $\psi^*\psi$ is given by

$$\frac{\partial}{\partial t}(\psi^*\psi) = -\frac{i\hbar}{2m_4}(\psi^*\nabla^2\psi - \psi\nabla^2\psi^*).$$ (2.17)

If we introduce the quantity

$$\mathbf{J}_P = -\frac{i\hbar}{2m_4}(\psi^*\nabla\psi - \psi\nabla\psi^*),$$ (2.18)

then eq. (2.17) becomes

$$\frac{\partial(\psi^*\psi)}{\partial t} = -\nabla\cdot\mathbf{J}_P.$$ (2.19)

In conventional quantum-mechanical language eq. (2.19) expresses conservation of probability, and $\mathbf{J}_P$ is the probability current density. However, when ψ is the superfluid wave function, $\psi^*\psi$ is the particle density and therefore $\mathbf{J}_P$ is the particle current density. Thus, eq. (2.19) ensures that when the number of particles in a given volume element decreases, there is a compensating net current out of the element.

We have already related the superfluid wave function to the superfluid density

$$\psi^*\psi = \psi_0^2 = \rho_s/m_4.$$ (2.3)

In the same way, the superfluid current density $\mathbf{j}_s$ can be related to $\mathbf{J}_P$. If, additionally, the explicit form of ψ (2.1) is substituted into eq. (2.18), we obtain

$$\mathbf{j}_s = m_4\mathbf{J}_P = \hbar\psi_0^2\nabla S.$$ (2.20)

In § 1.4 we saw that the superfluid velocity $\mathbf{v}_s$ is proportional to the gradient of the phase

$$\mathbf{v}_s = \frac{\hbar}{m_4}\nabla S.$$ (2.21)

Thus, from eqs. (2.3), (2.20) and (2.21), we find that

$$\mathbf{j}_s = \rho_s \mathbf{v}_s \tag{2.22}$$

as it should be. Finally we can write eq. (2.19) in the form

$$\frac{\partial \rho_s}{\partial t} = -\nabla \cdot \mathbf{j}_s, \tag{2.23}$$

which is the equation of continuity for the superfluid, an equation satisfied by all fluids in the absence of sources and sinks.

It is worth remarking that the use of m_4, the mass of a single atom, in eqs. (2.3), (2.10) and (2.21) is a convention. Any arbitrary mass could be employed since any scale factor could be absorbed into the phase S. The only requirement is that the same mass must be used in all three equations, because then the equation for $\mathbf{j}_s$ (2.22) and the continuity equation (2.23) are unaffected. Thus, eq. (2.21) should be regarded as merely a working definition of the superfluid velocity $\mathbf{v}_s$. When it comes to experimental measurements, $\mathbf{j}_s$ is the quantity normally observed. If ρ_s is measured at the same time, then values of $\mathbf{v}_s$ can be deduced from eq. (2.22), but $\mathbf{v}_s$ itself is not directly measurable.

Despite these reservations, the superfluid velocity is a useful concept. We shall now find the equation of motion for $\mathbf{v}_s$, starting from the basic equation (2.15). Taking the gradient of the latter, and using eq. (2.21) for ∇S, we find

$$m_4 \frac{\partial \mathbf{v}_s}{\partial t} = -\nabla(\mu + \tfrac{1}{2}m_4 v_s^2). \tag{2.24}$$

This is the equation of motion for the superfluid. We need to show that eq. (2.24) implies non-dissipative potential flow.

For that reason we introduce the operator D/Dt, known as the substantial derivative. Suppose that a fluid is moving with velocity $\mathbf{v}$, and that we consider an element of constant volume moving with the fluid. The total rate of change of some vector property of the fluid, $\mathbf{a}$ say, within the moving volume element is given by

$$\frac{D\mathbf{a}}{Dt} = \frac{\partial \mathbf{a}}{\partial t} + (\mathbf{v} \cdot \nabla)\mathbf{a} \tag{2.25}$$

where, for example, the x-component of $(\mathbf{v} \cdot \nabla)\mathbf{a}$ written out in full is

$$[(\mathbf{v} \cdot \nabla)\mathbf{a}]_x = v_x \frac{\partial \mathbf{a}_x}{\partial x} + v_y \frac{\partial \mathbf{a}_x}{\partial y} + v_z \frac{\partial \mathbf{a}_x}{\partial z}. \tag{2.26}$$

Thus the total derivative $D\mathbf{a}/Dt$ is made up of two parts, $\partial \mathbf{a}/\partial t$, which is the rate of change of $\mathbf{a}$ at a fixed point in space, and $(\mathbf{v} \cdot \nabla)\mathbf{a}$, the rate of change of $\mathbf{a}$ due to the motion of the fluid element.

We consider now the following equation for the superfluid velocity:

$$\frac{D\mathbf{v}_s}{Dt} = \frac{\partial \mathbf{v}_s}{\partial t} + (\mathbf{v}_s \cdot \nabla)\mathbf{v}_s = -\frac{1}{m_4}\nabla\mu. \tag{2.27}$$

This is Euler's equation for an ideal fluid; that is, a fluid which has no viscosity. Since μ is the potential energy for the superfluid, $\nabla\mu/m_4$ is the force per unit mass;

also Dv_s/Dt is the acceleration of the superfluid. In other words, eq. (2.27) is a form of Newton's Second Law. Using the properties of the operator ∇, we can write eq. (2.27) in an alternative way:

$$\frac{Dv_s}{Dt} = \frac{\partial v_s}{\partial t} + \nabla(\tfrac{1}{2}v_s^2) - v_s \times (\text{curl } v_s) = -\frac{1}{m_4}\nabla\mu. \tag{2.28}$$

We now distinguish between two different situations. Firstly, when

$$\text{curl } v_s = 0, \tag{2.29}$$

eq. (2.28) is identical to eq. (2.24), the superfluid equation of motion derived from the postulated macroscopic wave function. In addition, we recall that condition (2.29) follows immediately from the definition (2.21) which makes v_s proportional to the gradient of the wave function's phase S (§ 1.4). Normally, the superfluid component can be treated as incompressible, and putting ρ_s equal to a constant reduces the continuity equation (2.23) to

$$\nabla \cdot v_s = 0. \tag{2.30}$$

If now we substitute in this equation for v_s from eq. (2.21), we find

$$\nabla^2 S = 0, \tag{2.31}$$

which is Laplace's equation. Hence we see that the phase S acts as the velocity potential for the superfluid. Equation (2.24) therefore describes irrotational potential flow of the superfluid.

Secondly, we can have curl $v_s \neq 0$, in which case rotational motion occurs, and the full form of Euler's equation is needed. We shall treat He II in rotation in Chapter 4, where we shall discuss both persistent rotational currents and vortex motion, again using the macroscopic wave function (2.1) as a starting point.

Our next step is to write Euler's equation in its most useful form for the superfluid. To achieve that we find an expression for a small reversible change of chemical potential. The Gibbs energy for a fluid at rest is

$$G = U_0 - T\Sigma + pV. \tag{2.32}$$

Differentiation of this equation and substitution of eq. (2.5) for dU_0 gives

$$dG = \mu\, dN - \Sigma\, dT + V\, dp \tag{2.33}$$

from which it follows by integration that

$$N\mu(p,T) = G(N,p,T). \tag{2.34}$$

Thus we have an alternative definition of the chemical potential; at given values of pressure and temperature it is equal to the Gibbs energy per particle. In a system with fixed N, a change in μ can therefore be expressed as

$$N\, d\mu = \left(\frac{\partial G}{\partial p}\right)_{N,T} dp + \left(\frac{\partial G}{\partial T}\right)_{N,p} dT. \tag{2.35}$$

The differential coefficients are equal to V and $-\Sigma$ respectively, from eq. (2.33), whence eq. (2.35) becomes

$$N \, \mathrm{d}\mu = V \, \mathrm{d}p - \Sigma \, \mathrm{d}T, \tag{2.36}$$

indicating that variation of μ arises from variation of pressure and temperature. Division of eq. (2.36) throughout by $\mathrm{d}x$ say, yields the relation between gradients in the x-direction. There are equivalent equations for the y- and z-directions, and the three are collected in the one equation:

$$\nabla \mu = \frac{V}{N} \nabla p - \frac{\Sigma}{N} \nabla T. \tag{2.37}$$

Finally, we substitute for $\nabla \mu$ in eq. (2.27) and find

$$\frac{\mathrm{D}\mathbf{v}_s}{\mathrm{D}t} = \frac{\partial \mathbf{v}_s}{\partial t} + (\mathbf{v}_s \cdot \nabla)\mathbf{v}_s = -\frac{1}{\rho} \nabla p + \sigma \nabla T \tag{2.38}$$

where we have put $Nm_4/V = \rho$, the total density of the liquid and $\Sigma/Nm_4 = \sigma$, the total entropy per unit mass. We observe that Euler's equation links superfluid motion to both pressure and temperature gradients.

One important result follows immediately from eq. (2.38). When the superfluid flow is steady and irrotational,

$$\frac{\partial \mathbf{v}_s}{\partial t} = 0 = \nabla\left(\tfrac{1}{2}v_s^2 + \frac{p}{\rho} - \sigma T \right), \tag{2.39}$$

where we have used the fact that the whole liquid is incompressible by taking $1/\rho$ under the gradient sign. Integration of eq. (2.39) gives

$$\tfrac{1}{2}v_s^2 + \frac{p}{\rho} - \sigma T = \text{constant}, \tag{2.40}$$

which is Bernoulli's equation for potential flow of the superfluid. The constant takes the same value throughout a fluid undergoing potential flow. Bernoulli's equation is made more general in some cases by the addition of further terms; for instance, in applying it to a flowing helium film, the gravitational potential energy and the van der Waals potential are included (§2.7).

2.4   Motion of the normal fluid

In the last two sections we have ignored the normal fluid fraction of He II. We began from the hypothesis that a macroscopic wave function can be used to describe the superfluid, and arrived at Euler's equation (2.38). By itself this equation governs the motion of He II only in the limit as $\rho_n \to 0$. Now we must deal with the general case in which both components are present, each having its own velocity.

To avoid unnecessary complexity, we consider the situation in which the velocities are small. The superfluid velocity, $\mathbf{v}_s$, is much less than the velocity at which dissipation sets in, while the normal fluid velocity, $\mathbf{v}_n$, is well below that at which turbulence commences. Furthermore, there is no force of interaction between the two fluids. Thus the picture we use is of the superfluid in potential flow as before, and the normal fluid independently undergoing viscous flow (§1.1).

We can now write down the basic equations of the two-fluid model. The total density of He II is equal to the sum of the densities of the normal and superfluid components,

$$\rho = \rho_s + \rho_n. \tag{2.41}$$

The total mass current density is given by

$$\mathbf{j} = \rho_s \mathbf{v}_s + \rho_n \mathbf{v}_n. \tag{2.42}$$

Since the total mass of the fluid is conserved, ρ and $\mathbf{j}$ must obey the continuity equation

$$\frac{\partial \rho}{\partial t} = -\nabla \cdot \mathbf{j}. \tag{2.43}$$

The superfluid flow does not involve dissipation, so it is thermodynamically reversible. The normal fluid has viscosity, which causes dissipation, but only at a rate proportional to ∇v_n^2 (Landau and Lifshitz 1959). So long as $\mathbf{v}_n$ is small, this can be neglected and the flow of the normal fluid can be considered reversible as well. In these circumstances, the total entropy is conserved, and we can write a continuity equation involving the entropy density $\rho\sigma$. We saw in §1.1 that the entropy of He II is carried by the normal fluid, and so the corresponding entropy current density is equal to $\rho\sigma\mathbf{v}_n$, whence we have

$$\frac{\partial(\rho\sigma)}{\partial t} = -\nabla \cdot (\rho\sigma\mathbf{v}_n). \tag{2.44}$$

We shall now find equations of motion for incompressible flow, in which ρ, ρ_s, ρ_n and σ are all constant. From eq. (2.44) it follows that $\nabla \cdot \mathbf{v}_n = 0$, and thence from eqs. (2.42) and (2.43) that $\nabla \cdot \mathbf{v}_s = 0$. To describe the motion of the whole fluid we use a Navier–Stokes equation

$$\rho_s \left\{ \frac{\partial \mathbf{v}_s}{\partial t} + (\mathbf{v}_s \cdot \nabla)\mathbf{v}_s \right\} + \rho_n \left\{ \frac{\partial \mathbf{v}_n}{\partial t} + (\mathbf{v}_n \cdot \nabla)\mathbf{v}_n \right\} = -\nabla p + \eta_n \nabla^2 \mathbf{v}_n, \tag{2.45}$$

which expresses the fact that an external pressure gradient can cause motion of either or both components, and that resistance to flow comes only from the viscosity of the normal fluid, η_n being the viscosity coefficient. Equation (2.45) is derived from the condition for conservation of momentum of the whole fluid. However, the calculation is lengthy and will not be given here; a clear account of it, and other aspects of the two-fluid equations, can be found in Landau and Lifshitz (1959), Chapter 16.

Euler's equation for the superfluid (2.38), when multiplied by ρ_s, becomes

$$\rho_s \left\{ \frac{\partial \mathbf{v}_s}{\partial t} + (\mathbf{v}_s \cdot \nabla)\mathbf{v}_s \right\} = -\frac{\rho_s}{\rho} \nabla p + \rho_s \sigma \nabla T. \tag{2.46}$$

When this is subtracted from eq. (2.45), we are left with the following equation for the normal fluid:

$$\rho_n \left\{ \frac{\partial \mathbf{v}_n}{\partial t} + (\mathbf{v}_n \cdot \nabla)\mathbf{v}_n \right\} = -\frac{\rho_n}{\rho} \nabla p - \rho_s \sigma \nabla T + \eta_n \nabla^2 \mathbf{v}_n, \qquad (2.47)$$

an equation of the Navier–Stokes type, but with the addition of a thermal-gradient term. As we have stated before, these equations apply in the low-velocity limit when both components undergo incompressible flow. In order to make use of them in describing experimental situations we need to know the boundary conditions satisfied by $\mathbf{v}_s$ and $\mathbf{v}_n$.

Firstly we consider what happens at a solid wall in contact with He II. Since the normal fluid is a viscous liquid, both its normal and tangential components are zero. The normal component of $\mathbf{v}_s$ is also zero, but because the superfluid is non-viscous it can slip past the wall and thus the parallel component of $\mathbf{v}_s$ can take any value. The boundary conditions are modified in the particular case that heat is transferred between the liquid and the wall; obviously then the normal component of $\mathbf{v}_n$ will take a finite value, and there will be a compensating counterflow of superfluid to maintain constant total density. Apart from these conditions, we must also take into account the behaviour of ρ_s near the wall. Because the superfluid wave function must vanish at a solid boundary, it follows that $\rho_s \to 0$ smoothly as the wall is approached. A sudden jump in ρ_s at the wall is forbidden, because it would violate the requirement that ψ should be everywhere single-valued. However, the typical distance over which ρ_s falls to zero turns out to be quite small, of the order of the interatomic spacing (§ 6.8); consequently, as far as the macroscopic viewpoint of hydrodynamics is concerned, we still regard $\mathbf{v}_s$ as finite 'at the wall', even though the superfluid density is zero there.

Secondly, we consider the free surface of He II. Again, it is obvious that the normal components of both velocities will be zero. On the other hand, both tangential components are unrestricted in value. The boundary conditions satisfied by ψ and ρ_s at the free surface are still a matter for debate. It would be a reasonable guess that $\rho_s \to 0$, because $\psi = 0$ in the vapour, but the free surface does not provide a clear-cut division between the phases and therefore its influence on the superfluid is not well understood.

2.5 Generalized two-fluid model

The equations of motion in their basic form, (2.46) and (2.47), are adequate so long as the fluid velocities are small. As the velocities increase, many other factors, particularly those involving dissipation, come into play. To describe these, various non-linear terms are added to the equations, which become rather unwieldy. The origin of one of the extra terms is the dependence of the thermodynamic parameters of He II upon the relative velocity $(\mathbf{v}_n - \mathbf{v}_s)$ between the two fluids, as we shall now discuss.

Analysis of the situation in which the two fluids have different velocities is made simpler by the use of three coordinate frames, the laboratory frame which we call

K, a frame which moves with the superfluid, K_1, and a frame moving with the normal fluid, K_2. The velocities $\mathbf{v}_s$ and $\mathbf{v}_n$ that we have been using are measured in the frame K; thus K_1 moves with velocity $\mathbf{v}_s$ relative to K; and K_2 moves relative to K_1 with velocity $\mathbf{V} = \mathbf{v}_n - \mathbf{v}_s$. We treat the case in which $\mathbf{v}_s$ and $\mathbf{v}_n$ are constant in space and time, and establish the rules of transformation for energy and momentum between frames K_1 and K_2 (Fig. 2.1). Consider a free particle of mass M

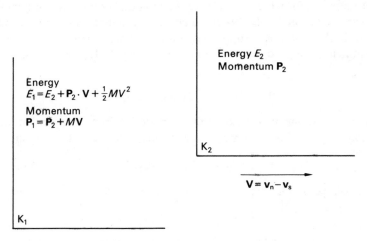

Energy
$$E_1 = E_2 + \mathbf{P}_2 \cdot \mathbf{V} + \tfrac{1}{2}MV^2$$

Momentum
$$P_1 = P_2 + M\mathbf{V}$$

Energy E_2
Momentum $\mathbf{P}_2$

K_2

$\mathbf{V} = \mathbf{v}_n - \mathbf{v}_s$

K_1

FIG. 2.1   Galilean transformation. Frame K_1 moves with the superfluid, K_2 with the normal fluid. Transformation shown for body of mass M having energy E_2 and momentum $\mathbf{P}_2$ in K_2.

having momentum $\mathbf{P}_2$ and energy E_2 in frame K_2. Its momentum in K_1 is

$$\mathbf{P}_1 = \mathbf{P}_2 + M\mathbf{V} \tag{2.48}$$

and its energy

$$E_1 = \frac{P_1^2}{2M} = \frac{P_2^2}{2M} + \mathbf{P}_2 \cdot \mathbf{V} + \tfrac{1}{2}MV^2 \tag{2.49}$$

or

$$E_1 = E_2 + \mathbf{P}_2 \cdot \mathbf{V} + \tfrac{1}{2}MV^2. \tag{2.50}$$

Equations (2.48) and (2.50) are the Galilean transformations; they are not restricted to free particles, and can be applied as they stand to any system of mass M possessing well-defined energy and momentum.

We now apply the transformations to the normal fluid. As in § 1.5 we treat the normal fluid as a gas of thermal excitations of energy ε and momentum p, whose energy spectrum $\varepsilon(p)$ is the Landau curve (Fig. 1.30). An essential rider is that ε and p must be measured relative to the superfluid component at rest, viz. the discussion leading to the Landau criterion in § 1.5: in other words, $\varepsilon(p)$ is the energy spectrum in frame K_1. To calculate the thermodynamic parameters we need the appropriate distribution function for the excitations. The total number of excitations is not conserved, so we use the Planck function

$$n(\varepsilon_2) = \{ \exp (\varepsilon_2/k_B T) - 1 \}^{-1}, \tag{2.51}$$

where the excitation energies ε_2 are measured in the frame K_2, the normal-fluid rest frame. Furthermore, calculations involving eq. (2.51) include integration over momentum space, for which we require the relation between ε_2 and the momentum of the excitation. To find that we consider the motion of the normal fluid relative to both K_1 and K_2. Suppose that the total mass of normal fluid is M_n; in K_2 its momentum is zero, and we call its energy E_2. In K_1, using eqs. (2.48) and (2.50),

$$\mathbf{P}_1 = M_n \mathbf{V}, \qquad E_1 = E_2 + \tfrac{1}{2} M_n V^2. \tag{2.52}$$

Now suppose that one further excitation is created with energy ε_2 and momentum $\mathbf{P}_2$ in frame K_2. The total momentum and energy in frame K_2 become

$$\mathbf{P}'_2 = \mathbf{p}_2, E'_2 = E_2 + \varepsilon_2. \tag{2.53}$$

In frame K_1, we find

$$\mathbf{P}'_1 = M_n \mathbf{V} + \mathbf{p}_2, E'_1 = E_2 + \tfrac{1}{2} M_n V^2 + \varepsilon_2 + \mathbf{p}_2 \cdot \mathbf{V}. \tag{2.54}$$

A comparison of eq. (2.54) with eq. (2.52) shows that the appearance of one excitation $(\varepsilon_2, \mathbf{p}_2)$ in frame K_2, involves additional momentum $\mathbf{p}_2$ and energy $\varepsilon_2 + \mathbf{p}_2 \cdot \mathbf{V}$ in frame K_1, the superfluid rest frame. Consequently we can put $\mathbf{p} = \mathbf{p}_2$ and

$$\varepsilon_2 = \varepsilon(p) - \mathbf{p} \cdot \mathbf{V}. \tag{2.55}$$

We can now write the distribution function (2.51) as

$$n(\varepsilon) = \left[\exp \left\{ \frac{\varepsilon(p) - \mathbf{p} \cdot (\mathbf{v}_n - \mathbf{v}_s)}{k_B T} \right\} - 1 \right]^{-1}. \tag{2.56}$$

For this reason, the relative velocity $(\mathbf{v}_n - \mathbf{v}_s)$ is an extra thermodynamic parameter for He II.

In terms of the Landau model, $(\mathbf{v}_n - \mathbf{v}_s)$ is the drift velocity of the excitation gas relative to the superfluid. Thus the drift momentum density of the excitations in frame K_1 is

$$\mathbf{j}_1 = \rho_n(\mathbf{v}_n - \mathbf{v}_s). \tag{2.57}$$

The total energy density u_1 measured in K_1 obeys the identity

$$du_1 = T d(\rho\sigma) + (\mu/m_4)d\rho + (\mathbf{v}_n - \mathbf{v}_s) \cdot d\mathbf{j}_1, \tag{2.58}$$

where the first two terms are contributed by the fluid at rest, as in eq. (2.5), and the last term is due to the relative motion of the two fluids.

All the necessary thermodynamic parameters can be obtained from eq. (2.58), by somewhat lengthy calculations. As an example, we have set one of these as a problem (2.3); we quote the result for the chemical potential (Landau and Lifshitz 1959, Chapter 16):

$$\frac{1}{m_4}\mu(p,T,\mathbf{v}_n-\mathbf{v}_s) = \frac{1}{m_4}\mu(p,T,0) - \frac{\rho_n}{2\rho}(\mathbf{v}_n-\mathbf{v}_s)^2, \tag{2.59}$$

where $\mu(p,T,0)$ is the chemical potential when there is no relative motion between the two fluids. Earlier we found an expression for $\nabla\mu(p,T,0)$ in eq. (2.37), which was then used to obtain the superfluid equation of motion (2.46). Now we use eq. (2.59) to substitute in eq. (2.28) for μ, giving

$$\rho_s\frac{D\mathbf{v}_s}{Dt} = -\frac{\rho_s}{\rho}\nabla p + \rho_s\sigma\nabla T + \frac{\rho_s\rho_n}{2\rho}\nabla(\mathbf{v}_n-\mathbf{v}_s)^2, \tag{2.60}$$

and when this is subtracted from eq. (2.45), we find

$$\rho_n\frac{D\mathbf{v}_n}{Dt} = -\frac{\rho_n}{\rho}\nabla p - \rho_s\sigma\nabla T - \frac{\rho_s\rho_n}{2\rho}\nabla(\mathbf{v}_n-\mathbf{v}_s)^2 + \eta_n\nabla^2\mathbf{v}_n. \tag{2.61}$$

The flow of He II at high velocities presents a complex experimental situation, in which it is difficult to distinguish between the many contributions to the dissipation of energy. However, extension of the two-fluid model gives some indication of the physical processes which can occur. For instance, one can define a critical value of $\mathbf{v}_n$ above which turbulence of the normal fluid begins, and a critical value of $(\mathbf{v}_n-\mathbf{v}_s)$ which marks the onset of a force of interaction between the two fluids, known as mutual friction. The most interesting consideration, however, is the breakdown of superfluidity and the factors that cause it. As yet there is no completely satisfactory explanation for the limitation of $\mathbf{v}_s$, but we shall discuss the more promising suggestions later in the book. For a summary of experimental work on He II at high velocities, the reader is referred to the book by Keller (1969), Chapter 8.

In addition, there have been several attempts to derive generalized forms of eqs. (2.60) and (2.61), the most significant being that of Bekarevich and Khalatnikov (1961), who included the motion of vortex lines in their treatment. An excellent discussion of this work can be found in Donnelly (1967), Chapter 4.

2.6 The flow of Helium II in channels

Some of the basic experiments which support the two-fluid model have been described in Chapter 1, notably Andronikashvili's experiment, which gave a direct measurement of ρ_n/ρ as a function of temperature, and the fountain effect. In the remainder of this chapter we shall review several more experiments, indicating how these relate to the two-fluid equations discussed earlier, and noting the limitations of the model.

A clear demonstration that the superfluid does undergo potential flow (eq. 2.24) has been given by the 'superfluid wind-tunnel' experiment of Craig and Pellam

(1957). The apparatus consists of a cylindrical vessel, illustrated in Fig. 2.2, immersed in a He II bath. The ends of the cylinder are superleaks, barriers which allow the passage of superfluid but which are impermeable to the normal component. Thus, when the heater is switched on, only superfluid flows through the wind-tunnel R; normal fluid is present in R as well, but it is stationary. After emerging from the upper superleak, the superfluid flows along the tube T, where it is mixed with normal fluid. The fluid velocity in T is measured with a Venturi meter, and from this, the superfluid velocity through R can be deduced. The nature of the flow within the wind-tunnel is indicated by the propeller which consists of two thin mica vanes suspended by a fine quartz fibre. A viscous liquid flowing past would exert a torque on the propeller because of turbulence round the vanes; this is the same effect as that experienced by an aerofoil. The torque would still be produced even if the viscosity were vanishingly small. In Fig. 2.3 we show the results for the superfluid, together with the estimated effect of a viscous liquid of the same density. The superfluid produces a torque down to a certain limiting velocity, below which no deflection of the propeller is detected. There appear to be two regimes of superflow in this experiment: at very low velocities the superfluid undergoes potential (irrotational) flow and, therefore, exerts no torque on the propeller; above a certain 'critical velocity', $\sim 6 \, \mathrm{mm \, s^{-1}}$ in this case, it behaves more like a classical liquid, and we suppose that turbulence occurs in the superfluid at these higher velocities.

Two points about the experiment deserve comment. The normal component is clamped and unable to move in the wind-tunnel, ensuring that the effects of superflow can be studied by themselves. This is particularly desirable when one is trying to disentangle the normal and superfluid properties of He II, and seeking the factors which lead to the breakdown of superfluidity. In many of the other experiments we describe in this chapter, we shall find that the normal fluid is held stationary in some way. The second point is the inference that a 'critical velocity' exists above which superfluid flow without dissipation of energy is impossible. In Chapter 1, we introduced the Landau criterion for superfluidity, which involved a critical velocity for the creation of individual excitations; for rotons this was $58 \, \mathrm{m \, s^{-1}}$ (eq. 1.65). This value is four orders of magnitude greater than the 'critical velocity' observed in the wind-tunnel, and clearly some other mechanism is needed to explain this low value.

Experiments to measure the viscosity of the normal fluid (eq. 2.61) were mentioned in §1.1. These involved the detection of the viscous drag on a solid body moving in the liquid. Methods involving Poiseuille flow through wide channels are not practicable, because the ease with which the superfluid moves, and the fact that its motion is independent of the pressure head, mask the normal fluid's motion.

A number of workers have examined superfluid flow in channels narrow enough for the normal fluid to be clamped because of its viscosity. As representative of this work we take the study of gravitational flow through superleaks by de Bruyn Ouboter et al. (1967). One form of the apparatus is shown in Fig. 2.4: the superleak is made by drilling a hole through a brass block and packing it with

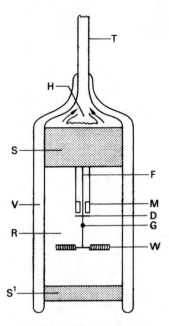

FIG. 2.2 Superfluid wind-tunnel (Craig and Pellam 1957). V—vacuum space, S, S'—superleaks, R—superfluid channel, W—propeller, F—quartz fibre, G—mirror; M—magnet, and D—copper disk for damping; H—heater, T—exit tube to Venturi gauge.

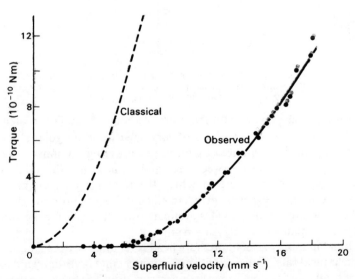

FIG. 2.3 Torque on propeller in superfluid wind-tunnel as a function of velocity of superfluid. For comparison, torque to be expected from viscous liquid of same density is shown by broken curve. (Craig and Pellam, 1957.)

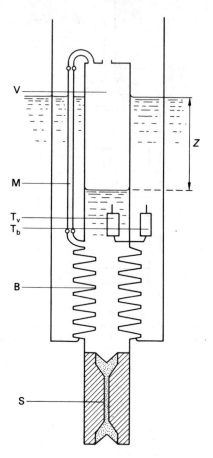

FIG. 2.4   Apparatus for measuring gravitational flow through superleaks (de Bruyn Ouboter *et al.*, 1967). S—superleak, B—bellows, M—side-arm to show level in V, V—superfluid reservoir, Z—pressure head, T_v, T_b—thermometers.

jeweller's rouge, a fine powder with grain size of about $0.5\,\mu$m. The result is a large number of tortuous, narrow channels of varying size and irregular shape. Flow through the superleak is initiated by raising or lowering the inner reservoir to produce a pressure difference between itself and the helium bath, and flow rate is measured by observing the rate at which the reservoir level changes. Other superleaks used in these experiments were made of Vycor, a porous glass with a sponge-like structure which provides even narrower channels for the superfluid, the average pore diameter being only 6 nm.

It was found that only a small pressure head is required to cause the superfluid to flow at a 'critical rate'. Once established, a critical rate is virtually independent of pressure head, but is a function of temperature. Fig. 2.5 shows the variation of critical flow rate with temperature for two different Vycor samples and three different jeweller's rouge superleaks. In all cases except the largest pore size, the

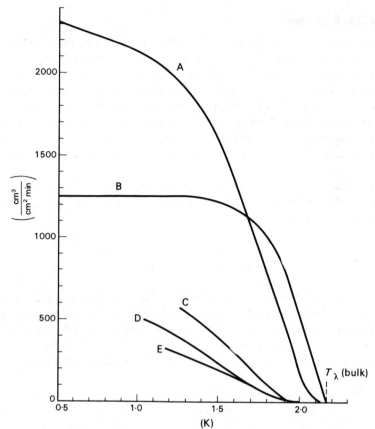

FIG. 2.5   Critical transfer rate as a function of temperature for gravitational flow through various superleaks (de Bruyn Ouboter *et al.*, 1967). A, B, D—jeweller's rouge, C, E—Vycor.

onset temperature for superflow is depressed below the bulk liquid value T_λ.

By 'critical rate' we mean that value of superfluid flow rate at which dissipation sets in. In these experiments, it was found that the region in which dissipation occurs is localized; this was accomplished by attaching three standpipes at different points of the superleak (Fig. 2.6) to monitor the variation of pressure. Since the liquid is almost stationary in each one of the standpipes, the liquid level indicates the pressure at the point of attachment. When the reservoir is emptying, or filling, at the critical rate, the entire pressure drop occurs between points A and B, that is, where the superfluid flow path is most constricted. On the other hand, when superfluid flow is maintained at a sub-critical rate by moving the reservoir, there is no pressure variation along the superleak. The interpretation of these results is that dissipation occurs in the region where there is a finite pressure gradient.

The superfluid equation (2.60) suggests that in a region where $\nabla p \neq 0$, the superfluid would be accelerated at a rate depending upon ∇p. However, it is clear that this is not what happens in the superleak, where a critical flow rate is reached which is independent of the pressure head. To account for this, eq. (2.60) needs some additional term to describe the dissipation of energy which occurs.

2.7 Liquid helium films

A film of liquid helium, typically about 100 atoms or 30 nm thick, forms on a solid surface in contact with He vapour at its saturation pressure. The existence of such a film is not remarkable in itself, but, seeing that it is much wider than the Vycor pores mentioned in the previous section, it is easily able to sustain superflow. On the other hand, the film thickness is sufficiently small to prevent the normal component from flowing almost entirely, with the result that a film is another kind of superleak. Indeed, the presence of a helium film on the wall of a vessel may lead to a serious loss of liquid, a possibility to be guarded against in the design of apparatus.

The properties of helium films are conveniently discussed in terms of the chemical potential. When the liquid in the film is at rest, the chemical potential per unit mass can be expressed as

$$\frac{1}{m_4}\mu = \frac{p}{\rho} - \sigma T + gz - \frac{\alpha}{y^3}, \tag{2.62}$$

where the pressure and temperature terms are present as before (eq. 2.37), gz is the gravitational potential energy per unit mass at height z and $-\alpha/y^3$ is the potential energy at a distance y from the wall due to the van der Waals forces, α being a constant whose value is determined by the material of the wall. Under isothermal, isobaric conditions, it follows from eq. (2.62) that the profile of a static helium film is

$$d = (\alpha/gH)^{1/3}, \tag{2.63}$$

where d is the film thickness at height H above the bulk liquid level; we have used the fact that the chemical potential has the same value at all points of the free surface. Direct measurement of film thicknesses have been accomplished by several methods, notably the use of polarized light (Jackson and Grimes 1958), and these lend some support to the profile (eq. 2.63). However, it is possible that further terms are needed in eq. (2.62), and values of α are not known with great accuracy.

When the film is in motion we must take account of eq. (2.59); in that case a chemical potential gradient is given by

$$\frac{1}{m_4}\nabla\mu = \frac{1}{\rho}\nabla p - \sigma\nabla T - \frac{\rho_n}{2\rho}\nabla(v_n - v_s)^2 + \nabla\left(gz - \frac{\alpha}{y^3}\right). \tag{2.64}$$

When this is substituted in the superfluid equation of motion (eq. 2.24), we obtain for steady-state conditions ($\partial v_s/\partial t = 0$) and constant pressure and temperature

$$\nabla\left(\frac{1}{2}\frac{\rho_s}{\rho}v_s^2 + gz - \frac{\alpha}{y^3}\right) = 0. \tag{2.65}$$

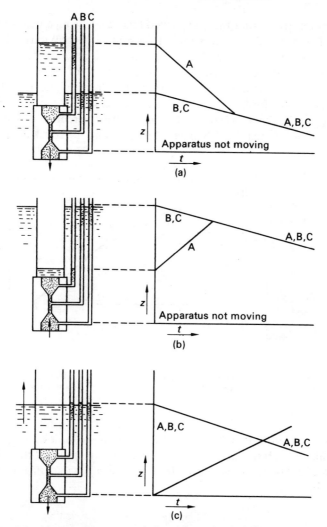

FIG. 2.6 Apparatus to measure pressure variation along a superleak (de Bruyn Ouboter *et al.*, 1967). A, B and C are manometers. (a) Reservoir emptying at critical flow rate, levels B and C are equal to each other and bath; (b) Reservoir filling at critical rate, levels B and C are again equal to each other and the bath; (c) Apparatus is raised at steady speed to ensure flow through superleak is at less than critical rate, A, B and C all level.

Integration of this equation leads to a modified form of Bernoulli's equation (cf. eqs. 2.39 and 2.40). Furthermore, by integrating eq. (2.65) along the film surface, we obtain a new profile equation, which indicates that a film through which the superfluid is moving should be thinner than the stationary film (Kontorovitch, 1956, Tilley, 1964). On the other hand, experiments in which film thickness was measured by a capacitative technique failed to find the predicted change of thickness (Keller, 1971).

In § 2.3 we saw that $\nabla\mu$ acts as the effective force on the superfluid; eq. (2.64) therefore demonstrates that there are several factors which can cause superfluid motion in a film. Of these, the simplest to study is a gravitational potential difference between the ends of a film, and this can be done, as we described in § 1.1, by observing the filling and emptying of a beaker. An informative series of experiments of this type has been carried out by Hammel and Keller and their co-workers; the basic form of their apparatus (Keller and Hammel 1966) is shown in Fig. 2.7, a glass beaker consisting of two parts with different cross-sections and a side-arm which joins the beaker between the rim A and the construction B. When

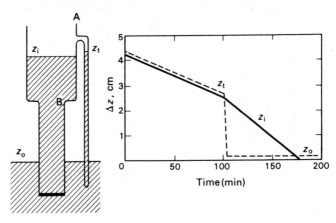

FIG. 2.7 Apparatus to probe variation of chemical potential along walls of beaker emptying and filling by film flow. Graph shows typical behaviour of levels during emptying. (Keller and Hammel, 1966.)

the inner level of liquid (height z_i) is higher than the outer level (z_o), the beaker empties by film flow. If the temperature and pressure are uniform, the difference in chemical potential between the two ends of the film is just the difference per atom of potential energy,

$$\Delta\mu = m_4 g(z_i - z_o),\qquad(2.66)$$

and the superfluid flows through the film in a manner analogous to an ordinary liquid in an inverted U-tube used as a siphon. The purpose of the side arm is to probe the variation of μ along the film; its diameter is small, one-tenth that of the narrower section, in order that the liquid level in the side arm (z_t) can adjust itself rapidly as flow conditions change.

The behaviour of the levels in an emptying experiment is shown in Fig. 2.7. While the inner level is between the probe and the constriction z_t remains level with z_i, except for a small difference due to capillarity. As soon as z_i falls below B, however, the level in the side arm drops rapidly to become equal to z_o. The level in a second probe, attached to the outside wall of the beaker (not shown in Fig. 2.7), remains fixed at the level of z_o. These observations indicate that the chemical potential varies along the film as shown in Fig. 2.8. With z_i above B, there is no

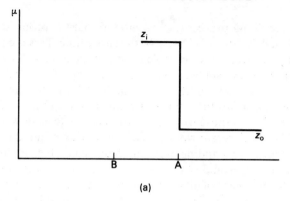

(a)

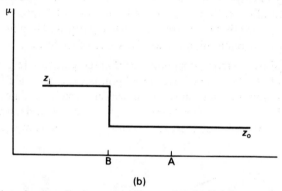

(b)

FIG. 2.8   Variation of μ along film in emptying process (idealized). (a) Inner level between A and B; (b) Inner level below B. (After Keller and Hammel, 1966.)

difference in chemical potential between z_i and the probe; with z_i below B, the chemical potential is uniform from z_o to the probe. The implication is that in each case the entire $\Delta\mu$ is concentrated in a very small region; the film chemical potential is virtually a step function with the step at A while z_i is above B, and at B with z_i below the constriction.

While z_i is between A and B it falls at a constant rate, indicating that the superfluid flow rate through the film is constant. This confirms that the flow rate is largely independent of the pressure head, as had been established by many workers previously (e.g., Daunt and Mendelssohn 1939). When the level passes the constriction, its rate of fall increases, but owing to the geometry of the beaker, this actually signifies a reduction of volume flow rate. Hence one infers that the flow rate is controlled at the smallest perimeter above the inner level, another result which had been established by Daunt and Mendelssohn (1939). What Keller and Hammel's experiment revealed is the coincidence of the region in which $\nabla\mu$ is non-zero with the part of the film which governs the flow rate, that is, where the superfluid velocity reaches its critical value and dissipation occurs. For the emptying process this region seems to be a small fraction of the film path, but examination of the filling process $(z_o > z_i)$ showed that the probe level remained between the two bulk levels $(z_o > z_t > z_i)$ for the entire time. This is consistent with spatial variation of μ over a considerable length of the film, with the implication that the dissipation region spreads over the inner wall of the beaker.

There have been several studies of thermally driven film flow, in which a temperature gradient makes the dominant contribution to $\nabla \mu$ (eq. 2.64), and the results are very similar to those from gravitational flow (e.g., Bowers *et al.*, 1951, Liebenberg, 1971). Much recent work on superflow in films and channels has been directed towards an understanding of the mechanism involved in dissipation and the limitation of the superfluid velocity. For a long time, it was thought that the critical velocity for the onset of dissipation would be determined primarily by the geometry of the flow channel. However, there is now considerable experimental support for a thermal fluctuation theory of dissipation (Langer and Fisher, 1967). It is convenient to postpone discussion of this theory and its relation to the flow experiments until after we have introduced vortices in Chapter 4.

In passing, we record that there are many other features of film flow which make it a very complicated phenomenon; we list some of these here:

(a) Flow rates show a strong dependence upon temperature (e.g. Fig. 2.9, taken from Allen *et al.*, 1965). The lower set of points were obtained from measurements of the flow rate into and out of beakers which were never completely filled. The decrease in rate as T approaches T_λ can be understood in terms of the variation of ρ_s / ρ, but the rise in rates below 1 K is mysterious.

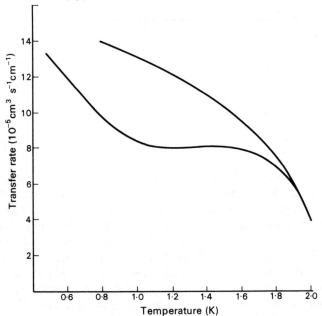

FIG. 2.9   Transfer rates as a function of temperature for emptying and filling of beaker by gravitational flow through helium films. Lower curve —rates measured in beaker where bulk liquid levels were kept at least 5 mm below rim. Upper curve —emptying rates measured after beaker had been totally immersed. (Allen *et al.*, 1965.)

(b) The upper set of points in Fig. 2.9 represent the flow rates out of beakers that had been immersed before measurement. This indicates that the history of the film may be important.

(c) Film rates have been observed to jump abruptly from one value to another (almost always lower) within one run at a fixed temperature (Harris-Lowe *et al.*, 1966; Allen and Armitage, 1967).

(d) It has been suggested that flow rates take 'quantized' values at a given temperature (Harris-Lowe *et al.*, 1966; Turkington and Edwards, 1968).

(e) Flow rates become dependent on the level difference to a greater extent below 1 K (Martin and Mendelssohn, 1971).

So far we have concentrated on saturated helium films; by reducing the vapour pressure below its saturation value, one can decrease the thickness of the film. These so-called unsaturated films are useful for the study of superfluidity in channels of restricted width. The variation of flow rate with temperature in unsaturated films (Fig. 2.10) is different from that in saturated films (Fig. 2.9). Furthermore, we find that the superflow onset temperature falls below T_λ and

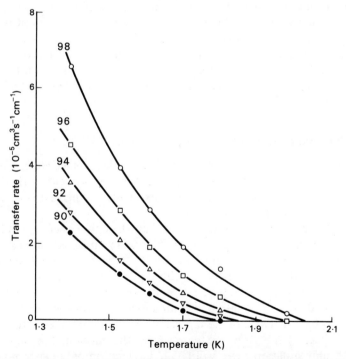

FIG. 2.10 Transfer rates for gravitational flow through unsaturated films of various thicknesses on glass substrate, plotted against temperature. Thicknesses decrease from right to left. Figures on curves indicate equilibrium pressure as percentage of saturation pressure. (After Brewer and Mendelssohn, 1961.)

continues to decrease as the thickness grows less (Fig. 2.11). An important point for inquiry is whether the cessation of superflow as the onset temperature is approached from below results from the superfluid density ρ_s tending to zero, or from the critical velocity for dissipation becoming zero. Experiments on *third sound* in unsaturated films have shown that the first alternative is ruled out (Kagiwada *et al.*, 1969). Third sound is a surface wave motion of the film, which we shall describe more fully in § 2.9; its velocity is proportional to $(\rho_s/\rho)^{1/2}$ and shows no sign of falling towards zero as the onset temperature is approached from below. The implication is that ρ_s remains finite through the transition in which superflow ceases. Thus, it seems probable that there is still superfluid present above the onset temperature but that it is incapable of flowing.

From the theoretical point of view, one can obtain at least a qualitative understanding of why the superfluid density should be reduced in a very thin film. In § 2.4 we discussed the boundary conditions which the superfluid wave function should satisfy. Its amplitude $(\rho_s/\rho)^{1/2}$ is zero at a solid wall and rises to the bulk value in a distance known as the *healing length*. At a free surface, the amplitude, or alternatively the gradient of the wave function, is zero. In a thin film, the wave-function amplitude would be less than the bulk value for a considerable fraction of the film thickness in order to comply with these boundary conditions (Fig. 2.12). When the thickness becomes of the same order as the healing length, $|\psi|$ and ρ_s do not reach the bulk value at all. Measurements on unsaturated films (Kagiwada *et al.*, 1969) have given very rough agreement with the theoretical estimates of the healing length (Josephson, 1966), but unfortunately they have not proved sensitive enough to decide between the alternative boundary conditions for the free surface. Incidentally, the healing length is the same as the vortex core radius which we shall meet in Chapter 4.

2.8   The fountain effect and heat transport in Helium II

In Chapter 1 we described the fountain effect (or thermomechanical effect) and noted that the transport of mass and heat are interdependent in He II. In this section we indicate briefly how experiments involving thermal gradients lend further support to the two-fluid equations.

We consider the arrangement of Fig. 2.13, in which two reservoirs of He II are connected by a channel of length Δx. In the first instance, we suppose that the channel is sufficiently narrow to prevent motion of the normal fluid, that is, it acts as a superleak. If the reservoirs begin with liquid at the same level, and a small temperature difference is established so that reservoir B is ΔT above A, we find that superfluid flows from A into B in an attempt to equalize the concentrations of the two fluids on either side of the superleak. The flow continues until a pressure excess Δp is reached in B, sufficient to prevent further motion of the superfluid. In this equilibrium situation there will be no acceleration of the superfluid in the channel, and so from eq. (2.60),

$$\frac{D\mathbf{v}_s}{Dt} = -\frac{1}{m_4}\frac{\Delta\mu}{\Delta x} = -\frac{1}{\rho}\frac{\Delta p}{\Delta x} + \sigma\frac{\Delta T}{\Delta x} = 0, \qquad (2.67)$$

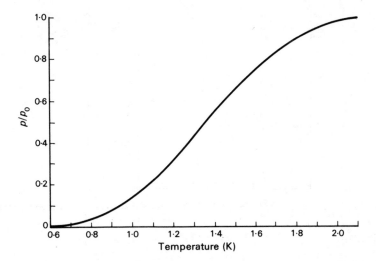

FIG. 2.11   Onset of superflow in unsaturated films. p—equilibrium pressure; p_0—saturation pressure. Smoothed curve from many experimental sources. (Goodstein and Elgin, 1969.)

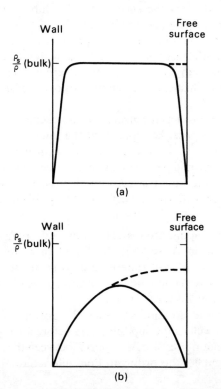

FIG. 2.12   Variation of wave-function amplitude, and ρ_s/ρ, in a helium film. (a) Film thickness much greater than healing length. (b) Film thickness of same order as healing length. Full lines, $\psi = 0$ at free surface; broken lines, $\nabla\psi = 0$ at free surface.

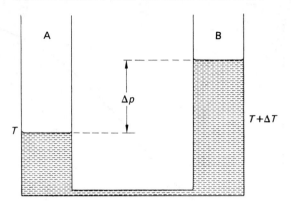

FIG. 2.13   Thermomechanical effect in He II. Two reservoirs connected by short channel.

which leads to

$$\frac{\Delta p}{\Delta T} = \rho\sigma, \tag{2.68}$$

the fountain effect equation first derived by H. London (1939). The entropy of He II can be calculated from the specific heat; a table of specific heats and entropies is given in the Appendix of the book by Wilks (1967). Using the entropy at $1\cdot3$ K, we see that the fountain effect is appreciable, for at this temperature a difference of only one millidegree produces a head of about 1 cm of He II between the two baths. The fountain equation has been verified in a number of experiments, all more or less sophisticated developments of the basic arrangement (Fig. 2.13).

The reverse process, known as the mechano-caloric effect, can also be described in terms of Fig. 2.13. Suppose the liquid is at the same level in both reservoirs, which are now enclosed in a constant-temperature bath at T. If an excess pressure is now applied to A, superfluid passes through the superleak into B, and it carries no entropy. However, the liquid already in B possesses entropy $\rho\sigma$ per unit volume; to bring the newly arrived superfluid into equilibrium an amount of heat

$$Q = \rho\sigma T \tag{2.69}$$

per unit volume must be supplied to vessel B. Conversely, the same amount of heat Q must be extracted from A, so that the liquid there remains at the same temperature. Thus the transfer of superfluid at constant temperature with the normal fluid stationary is accompanied by a flow of heat. This effect has been confirmed by Brewer and Edwards (1958).

Finally we consider the situation in which the normal fluid can flow in the connecting channel as well as the superfluid. As before a temperature difference ΔT gives rise to a fountain pressure $\Delta p = \rho\sigma\Delta T$ between the baths. The pressure head causes normal fluid flow according to Poiseuille's Law, the volume flow rate being

$$\dot{V}_n = \frac{\beta}{\eta_n}\frac{\Delta p}{\Delta x}, \tag{2.70}$$

where β is a geometrical factor. The normal component flows from the hotter reservoir (B) to the colder one (A), taking entropy with it at the rate of $\dot{V}_n \rho \sigma$, while the superfluid carrying no entropy flows in the opposite direction to maintain the constant total density. The result is a large heat flux down the temperature gradient, heat being transported at the rate

$$\dot{Q} = \frac{\beta T (\rho \sigma)^2}{\eta_n} \frac{\Delta T}{\Delta x}, \tag{2.71}$$

an equation first derived by F. London and Zilsel (1948). It has been verified by Brewer and Edwards (1959), who used their measured heat fluxes to calculate the normal viscosity coefficient; they found good agreement with the results of more direct methods, such as the rotation viscometer (§ 1.1).

2.9 The propagation of sound in Helium II

In the final section of this chapter we discuss the various kinds of sound waves which can be propagated in He II. Firstly, we derive the equations for first and second sound from the two-fluid equations, following Atkins (1959). We use the continuity equation (2.43) with (2.42):

$$\frac{\partial \rho}{\partial t} = -\nabla \cdot \mathbf{j} = -\nabla \cdot (\rho_s \mathbf{v}_s + \rho_n \mathbf{v}_n) \tag{2.72}$$

and the equation for entropy conservation (2.44)

$$\frac{\partial (\rho \sigma)}{\partial t} + \nabla \cdot (\rho \sigma \mathbf{v}_n) = 0. \tag{2.73}$$

From the equations of motion (2.60) and (2.61), we omit terms quadratic in the velocities

$$\rho_s \frac{\partial \mathbf{v}_s}{\partial t} = -\frac{\rho_s}{\rho} \nabla p + \rho_s \sigma \nabla T, \tag{2.74}$$

$$\rho_n \frac{\partial \mathbf{v}_n}{\partial t} = -\frac{\rho_n}{\rho} \nabla p - \rho_s \sigma \nabla T. \tag{2.75}$$

Thus we consider the case of low velocities and ignore irreversible factors. Adding eqs. (2.74) and (2.75) together we have

$$\frac{\partial \mathbf{j}}{\partial t} = -\nabla p, \tag{2.76}$$

where further second-order terms such as $\mathbf{v}_s (\partial \rho_s / \partial t)$ have been ignored. In combination eqs. (2.72) and (2.76) yield

$$\frac{\partial^2 \rho}{\partial t^2} = \nabla^2 p \tag{2.77}$$

the first of the two equations that we need. To obtain the second, we eliminate ∇p between eqs. (2.74) and (2.76),

$$\rho_n \frac{\partial}{\partial t}(\mathbf{v}_n - \mathbf{v}_s) = -\rho\sigma\nabla T,$$

(2.78)

which confirms that a temperature gradient produces relative motion between the two fluids, the situation discussed at the end of the previous section. Furthermore, from eqs. (2.72) and (2.73) we find that

$$\nabla \cdot (\mathbf{v}_n - \mathbf{v}_s) = -\frac{\rho}{\rho_s \sigma}\frac{\partial\sigma}{\partial t}$$

(2.79)

and this relation when combined with eq. (2.78) gives

$$\frac{\partial^2\sigma}{\partial t^2} = \frac{\rho_s}{\rho_n}\sigma^2\nabla^2 T.$$

(2.80)

It is now convenient to express both the pressure and the temperature as functions of entropy and density, whereupon eqs. (2.77) and (2.80) are re-expressed as

$$\frac{\partial^2\rho}{\partial t^2} = \left(\frac{\partial p}{\partial\rho}\right)_\sigma \nabla^2\rho + \left(\frac{\partial p}{\partial\sigma}\right)_\rho \nabla^2\sigma$$

(2.81)

and

$$\frac{\partial^2\sigma}{\partial t^2} = \frac{\rho_s}{\rho_n}\sigma^2\left[\left(\frac{\partial T}{\partial\rho}\right)_\sigma \nabla^2\rho + \left(\frac{\partial T}{\partial\sigma}\right)_\rho \nabla^2\sigma\right].$$

(2.82)

We look for plane-wave solutions of these two equations, that is, fluctuations of density and entropy,

$$\rho = \rho_0 + \rho'\exp\left[i\omega(t - z/u)\right],$$

$$\sigma = \sigma_0 + \sigma'\exp\left[i\omega(t - z/u)\right],$$

(2.83)

where u is the velocity in the z-direction, and ω is the angular frequency. Substituting eqs. (2.83) into eqs. (2.81) and (2.82), we find

$$\left\{\left(\frac{u}{u_1}\right)^2 - 1\right\}\rho' - \left(\frac{\partial p}{\partial\sigma}\right)_\rho\left(\frac{\partial\rho}{\partial p}\right)_\sigma \sigma' = 0,$$

$$\left(\frac{\partial T}{\partial\rho}\right)_\sigma\left(\frac{\partial\sigma}{\partial T}\right)_\rho \rho' - \left\{\left(\frac{u}{u_2}\right)^2 - 1\right\}\sigma' = 0,$$

(2.84)

where

$$u_1^2 = \left(\frac{\partial p}{\partial\rho}\right)_\sigma$$

(2.85)

and

$$u_2^2 = \frac{\rho_s}{\rho_n}\sigma^2\left(\frac{\partial T}{\partial\sigma}\right)_\rho.$$

(2.86)

The simultaneous equations (2.84) are mutually consistent provided that

$$\left\{\left(\frac{u}{u_1}\right)^2 - 1\right\}\left\{\left(\frac{u}{u_2}\right)^2 - 1\right\} = \left(\frac{\partial p}{\partial\sigma}\right)_\rho\left(\frac{\partial\rho}{\partial p}\right)_\sigma\left(\frac{\partial T}{\partial\rho}\right)_\sigma\left(\frac{\partial\sigma}{\partial T}\right)_\rho,$$

$$= \frac{C_p - C_V}{C_p} \simeq 0.$$

(2.87)

In the last equation the principal specific heats have been introduced by means of well-known thermodynamic relations; in fact, C_p and C_V are very nearly equal in He II and we can put the right-hand side of eq. (2.87) approximately equal to zero. The two (approximate) solutions of eq. (2.87) correspond to two different wave motions:

a) *First Sound*

$$u \simeq u_1; \qquad \sigma' = 0, \rho' \neq 0 \qquad \text{from eq. (2.84)},$$
$$\nabla T = 0 \qquad \text{from eq. (2.80)},$$
$$\mathbf{v}_n = \mathbf{v}_s \qquad \text{from eqs. (2.74) and (2.75)}.$$

These conditions describe the propagation of a density wave in the liquid at almost constant entropy, with the two fluids moving together; indeed eq. (2.85) gives the usual definition of the velocity of ordinary sound (*first sound*).

b) *Second Sound*

$$u \simeq u_2; \qquad \rho' = 0, \sigma' \neq 0 \qquad \text{from eq. (2.84)},$$
$$\nabla p = 0 \qquad \text{from eq. (2.80)},$$
$$\rho_s \mathbf{v}_s + \rho_n \mathbf{v}_n = 0 \qquad \text{from eq. (2.76)}.$$

These conditions describe the propagation of an entropy wave (or *temperature wave*) at almost constant total density; the last equation indicates that the two fluids move in antiphase in order to keep the net flow of mass at zero.

We reiterate that the equations from which these solutions have been obtained are approximations, from which all non-linear terms and irreversible effects have been omitted. In practice, first and second sound are almost always both present, and they are coupled together.

The velocity of first sound has been measured by determining the wavelength (van Itterbeek *et al.*, 1957) and by a radar-pulse technique, in which resonant quartz-crystal oscillators were used as transmitter and detector (Chase, 1953). Typical results for the velocity under the saturated vapour pressure are shown in Fig. 2.14.

Second sound has been studied by setting up standing waves in a resonator tube, the transmitter being an electrical heater supplied with alternating current, and the detector being a resistance thermometer (Peshkov, 1960). The second-sound frequency is twice that of the a.c. The variation of second-sound velocity with temperature is shown in Fig. 2.15. Incidentally, rewriting eq. (2.86) as

$$u_2^2 = \frac{\rho_s\, T\sigma^2}{\rho_n\, C_p},$$

we see that the second-sound velocity provides another way of measuring the parameters ρ_s and ρ_n.

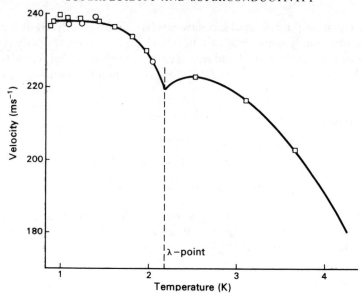

Fig. 2.14   The velocity of sound in liquid He⁴ under the saturated vapour pressure. ($\square$)—2·0 MHz, ($\odot$)—12·1 MHz. (Chase, 1953.)

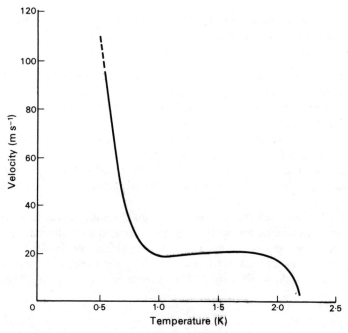

Fig. 2.15   Velocity of second sound in He II as a function of temperature. (Wilks 1967, after Peshkov, 1960.)

The attenuation of first sound in He II is caused by the usual classical factors of viscosity and thermal conductivity, which serve to dissipate the energy of the wave. In addition, certain relaxation effects of the thermal excitations contribute to the attenuation. These arise because the phonon and roton number densities require a finite time to adjust to the appropriate values when the liquid in a particular volume element undergoes compression or rarefaction. All the factors that attenuate first sound contribute also to the absorption of second-sound waves, together with an extra thermal conductivity due to the drift motion of the excitations. To take account of these numerous effects, the equations of the two-fluid model can be extended, by the inclusion of more terms. The solution of the final equations is then a complex matter; for their derivation and solution the reader is referred to the books by Khalatnikov (1965), and by Wilks (1967), who also describes the experimental verification of the theoretical predictions.

Finally, we refer briefly to two further wave motions in He II which are related to second sound. One of these, third sound, has already been mentioned in the discussion of unsaturated films in §2.7. Third sound is a surface wave on a He II film (Fig. 2.16), in which the superfluid oscillates parallel to the substrate, while the normal fluid is clamped because of its viscosity. Thus, at a wave crest, there is a

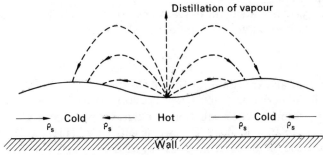

FIG. 2.16   A third-sound wave on a He II film.

larger fraction of superfluid than in equilibrium, and at a trough a smaller fraction, with the result that the temperature is raised in the trough but lowered on the crest. A third-sound wave is therefore accompanied by temperature fluctuations, although local temperature gradients are appreciably reduced by evaporation from the troughs and condensation at the crests. To obtain an approximate expression for the phase velocity of third sound, one can adapt the equation for the velocity of a surface wave on a shallow classical liquid of depth d

$$v_p^2 = \left(\frac{f\lambda}{2\pi} + \frac{2\pi\sigma_t}{\rho\lambda} \right) \tanh \frac{2\pi d}{\lambda}, \tag{2.88}$$

in which f is the force per unit mass due to the van der Waals attraction of the substrate, which is the main restoring force, and σ_t is the surface tension. For very long wavelengths, the first term is dominant, and taking account of the fact that only the superfluid component moves, the third-sound velocity can be written

$$u_3 = (\rho_s f d/\rho)^{1/2}. \tag{2.89}$$

In magnitude, u_3 is of the same order as the observed flow velocities in saturated films (10 to $100 \, \text{cm s}^{-1}$). Third sound can be excited by shining an infrared beam on the film and chopping it at the desired frequency, and the accompanying oscillations of film thickness can be detected by a polarized-light technique. A full treatment of the theory of third sound requires the use of the two-fluid equations with the appropriate boundary conditions. For a review of the experimental and theoretical aspects of third sound, see the article by Atkins and Rudnick (1970).

Fourth sound comprises the oscillation of superfluid in narrow rectangular channels, with the normal fluid once again stationary. There are temperature oscillations, and since there is no free surface, oscillations of total density as well (Shapiro and Rudnick, 1965).

Problems

2.1 Confirm that when the superfluid wave function (eq. 2.1) is substituted into Schrödinger's equation (2.10), one obtains eq. (2.12) and the continuity equation (2.23).

2.2 Show that eqs. (2.27) and (2.28) are equivalent forms of Euler's equation for the superfluid.

2.3 Show, by thermodynamical reasoning, that eq. (2.58) implies that the pressure according to the two-fluid model is given by

$$p = -u_1 + T\rho\sigma + \mu\rho + \rho_n(\mathbf{v}_n - \mathbf{v}_s)^2.$$

Hence, find an expression for the Gibbs free energy when there is relative motion, and derive eq. (2.59). (*Hint:* use eq. (2.34).)

2.4 Using the fountain-effect equation, calculate the pressure head in cm of He II equivalent to 10^{-3} K at 1·3 K. (Specific entropy of He II $= 0.085 \, \text{J K}^{-1} \, \text{g}^{-1}$; density of He II $= 0.145 \, \text{g cm}^{-3}$, both at 1·3 K).

2.5 Use the definitions of C_p and C_V to show that the right-hand side of eq. (2.87) can be written as $(C_p - C_V)/C_p$.

2.6 Show that the attenuation of a first-sound wave in He II due to the viscosity of the normal fluid, η_n, can be described by an attenuation coefficient $\alpha \propto \eta_n \omega^2 / \rho u_1^3$. (Add the term $\eta_n \nabla^2 \mathbf{v}_n$ to eq. (2.75), as in eq. (2.61), and look for a solution corresponding to a plane first-sound wave of frequency ω. Show that this leads to a complex expression for the velocity and calculate α from this, neglecting dispersion).

References

ALLEN, J. F. and ARMITAGE, J. G. M., 1967. *Proc. 10th Int. Conf. on Low Temp. Phys.*, Moscow 1966, vol. 1, p. 439 (Viniti, Moscow).

ALLEN, J. F., MATHESON, C. C. and WALKER, C. M., 1965, *Phys. Lett.* **19**, 199.

ANDERSON, P. W., 1966, *Quantum Fluids* (ed. D. F. Brewer), p. 146 (North–Holland, Amsterdam).

ATKINS, K. R., 1959, *Liquid Helium* (Cambridge University Press, London).

ATKINS, K. R. and RUDNICK, I., 1970, *Progress in Low Temp. Phys.*, vol. VI (ed. C. J. Gorter), ch. 2 (North–Holland, Amsterdam).

BEKAREVICH, I. L. and KHALATNIKOV, I. M., 1961, *Zh. Eksp. i Teor. Fiz.* **40**, 920. (Transl. in *Soviet Physics J E T P* **13**, 643 (1961)).
BOWERS, R., BREWER, D. F. and MENDELSSOHN, K., 1951, *Phil. Mag.* **42**, 1445.
BREWER, D. F. and EDWARDS, D. O., 1958, *Proc. Phys. Soc.* **71**, 117.
BREWER, D. F. and EDWARDS, D. O., 1959, *Proc. Roy. Soc.* **A251**, 247.
BREWER, D. F. and MENDELSSOHN, K., 1961, *Proc. Roy. Soc.* **A260**, 1.
CHASE, C. E., 1953, *Proc. Roy. Soc.* **A220**, 116.
COWLEY, R. A. and WOODS, A. D. B., 1971, *Can. J. Phys.* **49**, 177.
CRAIG, P. P. and PELLAM, J. R., 1957, *Phys. Rev.* **108**, 1109.
DAUNT, J. G. and MENDELSSOHN, K., 1939, *Proc. Roy. Soc.* **A170**, 423 and 439.
DE BRUYN OUBOTER, R., TACONIS, K. W. and VAN ALPHEN, W. M., 1967, *Progress in Low Temperature Physics*, vol. V (ed. C. J. Gorter), ch. 2 (North–Holland, Amsterdam).
DONNELLY, R. J., 1967, *Experimental Superfluidity* (University of Chicago Press).
GOODSTEIN, D. L. and ELGIN, R. L., 1969, *Phys. Rev. Lett.* **22**, 383.
HARRIS-LOWE, R. F., MATE, C. F., McCLOUD, K. L. and DAUNT, J. G., 1966, *Phys. Lett.* **20**, 126.
JACKSON, L. C. and GRIMES. L. G., 1958, *Adv. Phys.* **7**, 435.
JOSEPHSON, B. D., 1966, *Phys. Lett.* **21**, 608.
KAGIWADA, R. S., FRASER, J. C., RUDNICK, I. and BERGMAN, D., 1969, *Phys. Rev. Lett.* **22**, 338.
KELLER, W. E., 1969, *Helium-3 and Helium-4* (Plenum Press, New York).
KELLER, W. E. and HAMMEL, E. F., 1966, *Phys. Rev. Lett.* **17**, 998.
KELLER, W. E., 1971, *Proc. 12th Int. Conf. on Low Temp. Phys.*, Kyoto 1970, p. 125 (Academic Press of Japan, Tokyo).
KHALATNIKOV, I. M., 1965, *An Introduction to the Theory of Superfluidity* (Benjamin, New York).
KONTOROVITCH, V. M., 1956, *Zh. Eksp. i Teor. Fiz.* **30**, 805. (Transl. in *Soviet Physics J E T P* **3**, 770, (1956)).
LANDAU, L. D., 1941, *J. Phys. Moscow* **5**, 71.
LANDAU, L. D. and LIFSHITZ, E. M., 1959, *Fluid Dynamics* (Pergamon, Oxford).
LANGER, J. S. and FISHER, M. E., 1967, *Phys. Rev. Lett.* **19**, 560.
LIEBENBERG, D. H., 1971, *J. Low Temp. Phys.* **5**, 267.
LONDON, F. and ZILSEL, P. R., 1948, *Phys. Rev.* **74**, 1148.
LONDON, H., 1939, *Proc. Roy. Soc.* **A171**, 484.
MARTIN, D. J. and MENDELSSOHN, K., 1971, *J. Low Temp. Phys.* **5**, 211.
MARTIN, P. C., 1966, *Quantum Fluids* (ed. D. F. Brewer), p. 230 (North–Holland, Amsterdam).
PESHKOV, V. P., 1960, *Zh. Eksp. i Teor. Fiz.* **38**, 799. (Transl. in *Soviet Physics J E T P* **11**, 580, 1960).
PINES, D., 1961, *The Many Body Problem* (Benjamin, New York).
SHAPIRO, R. S. and RUDNICK, I., 1965, *Phys. Rev.* **137**, A1383.
TILLEY, J., 1964, *Proc. Phys. Soc.* **84**, 77.
TROUP, G. J., 1967, *Optical Coherence Theory— Recent Developments* (Methuen, London).
TURKINGTON, R. R. and EDWARDS, M. H., 1968, *Phys. Rev.* **168**, 160.
VAN ITTERBEEK, A., FORREZ, G. and TEIRLINEK, M., 1957, *Physica* **23**, 63 and 905.
WILKS, J., 1967, *The Properties of Liquid and Solid Helium* (Oxford University Press, London).

Chapter 3

Superconductors: Electrodynamics and Tunnelling

3.1 Introduction

In this chapter we are going to deal with two separate but related topics. First we shall discuss in somewhat more detail than in §1.2 some features of the response of superconductors to electromagnetic fields. The simplest response to a d.c. field is the exclusion of a magnetic field by screening currents flowing within a penetration depth λ of the surface. The nature of the equations describing this screening depends among other things on whether λ is larger or smaller than l, the mean free path in the normal state. This d.c. effect in a superconductor is analogous to the screening of microwave radiation by a normal metal. There it is found that the radiation is confined to a skin depth δ, and the relationship between the current density $\mathbf{J}$ and the microwave field $\mathbf{E}$ depends on whether δ is larger or smaller than l. Because of the importance of the analogy, we shall devote a preliminary section to the microwave skin effect in a normal metal.

We shall pass on from the d.c. screening in a superconductor to the principal a.c. effect, namely infrared absorption, which is governed by the energy gap 2Δ. This provides the link with the second main topic of this chapter, the tunnelling of quasiparticles through an insulating barrier between superconductors. We shall see that this is a very sensitive tool for studying the energy gap and further properties affecting the density of quasiparticle states.

3.2   Microwave skin effect in a normal metal

Suppose we have microwave radiation incident upon the surface of a metal, as in Fig. 3.1. For definiteness, we take the surface to be the xy-plane, and we deal

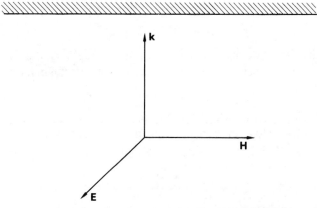

FIG. 3.1   Geometry for calculation of surface impedance.

with normal incidence, so that the wave is propagating in the z-direction. Finally, we restrict our attention to plane-polarized radiation, and take the x-axis along the direction of the $\mathbf{E}$-vector of the radiation, $\mathbf{E} = (E, 0, 0)$. The propagation in the metal is governed by Maxwell's equations:

$$\operatorname{curl} \mathbf{E} = -\partial \mathbf{B}/\partial t, \tag{3.1}$$

$$\operatorname{curl} \mathbf{H} = \mathbf{J} + \partial \mathbf{D}/\partial t. \tag{3.2}$$

To find a propagation equation for $\mathbf{E}$, we need a relation between $\mathbf{J}$ and $\mathbf{E}$; for the time being we suppose that this is given by Ohm's Law:

$$\mathbf{J} = \sigma \mathbf{E}. \tag{3.3}$$

We shall see later that Ohm's Law is adequate at low frequencies, but not always at high frequencies. If we eliminate $\mathbf{J}$ and the magnetic vectors between the three equations, we find the following equation for $\mathbf{E}$:

$$\frac{\partial^2 E}{\partial z^2} = \sigma \mu_0 \frac{\partial E}{\partial t} + \varepsilon \mu_0 \frac{\partial^2 E}{\partial t^2}, \tag{3.4}$$

in which the first term on the right-hand side comes from the conduction current $\mathbf{J}$, and the second from the displacement current $\partial \mathbf{D}/\partial t$. We are assuming that the metal is non-magnetic, with permeability μ_0. For a wave of angular frequency ω, the two terms in eq. (3.4) have the relative magnitudes $\sigma : \omega \varepsilon$, and for any normal metal at microwave frequency ($\omega \sim 10^{10}\,\mathrm{s}^{-1}$) one has $\sigma \gg \omega \varepsilon$. This means that we can drop the displacement current, and work with the simplified equation

$$\frac{\partial^2 E}{\partial z^2} = \sigma \mu_0 \frac{\partial E}{\partial t}. \tag{3.5}$$

To find a solution of eq. (3.5), we note that the propagation into the metal must be attenuated, because the Ohmic current removes energy from the wave. We work now with a wave of a single frequency ω, and take as a trial solution

$$E = E_0 \exp{(ikz - i\omega t)} \exp{(-z/\delta)}, \tag{3.6}$$

which is an attenuated plane wave. Substitution into eq. (3.5) gives

$$(ik - 1/\delta)^2 = -i\omega\sigma\mu_0. \tag{3.7}$$

Since the right-hand side of this equation is purely imaginary, the real part of the left-hand side must vanish, so that $k = 1/\delta$. Equating the imaginary parts, we find

$$\delta = (2/\sigma\mu_0\omega)^{1/2}. \tag{3.8}$$

The quantity that is usually measured is the surface impedance Z_s, which is the ratio of the electric field at the surface to the total current flowing across a unit line in the surface. That is,

$$Z_s = E_0 / \int_0^\infty \mathbf{J} \, dz. \tag{3.9}$$

Since $\mathbf{J}$ is $\sigma\mathbf{E}$, where $\mathbf{E}$ is given in eq. (3.6), we have

$$Z_s = (1/\delta - ik)/\sigma. \tag{3.10}$$

We recall that $k = 1/\delta$, so that the reactive and resistive parts of Z_s are equal. A typical experimental arrangement for surface impedance measurements is sketched in Fig. 3.2. The specimen is placed in a microwave resonant cavity; the reactive part of the specimen impedance shifts the frequency at which resonance

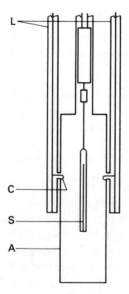

FIG. 3.2   Resonant cavity for surface-impedance measurements (after Dheer, 1961). The specimen S is the inner conductor of a coaxial resonator, outer walls A. The microwaves are fed in and out via the transmission lines L and the coupling loops C.

occurs, and the resistive part, which introduces extra dissipation, broadens the power absorption curve around the resonance frequency of the cavity.

Equation (3.8) gives the skin depth δ, derived on the assumption that Ohm's Law holds for the propagation into the metal. Note, however, that δ decreases as the frequency ω increases. At liquid-helium temperatures in a pure metal, a rough numerical estimate generally shows that at microwave frequencies δ is less than l, the mean free path of the electrons. We shall see that if δ is less than l, then $\mathbf{J}$ and $\mathbf{E}$ are in fact no longer related by Ohm's Law. The frequency region for which $\delta > l$ is called the *normal skin-effect* region, and that for which $\delta < l$ is the *anomalous skin-effect* region. We now turn to the latter region, and see how the theory we have outlined needs to be modified there.

If we write Ohm's Law as $\mathbf{J}(\mathbf{r}) = \sigma \mathbf{E}(\mathbf{r})$, we see that it assumes that the value of $\mathbf{J}$ at the point $\mathbf{r}$ is determined simply by the value of $\mathbf{E}$ at the same point $\mathbf{r}$. However, in the case of the anomalous skin effect, when we have $\delta < l$, then the electric field varies substantially along the path of an electron between successive collisions.

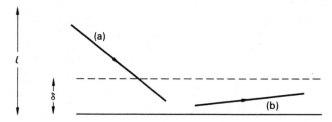

FIG. 3.3   Electron trajectories between collisions in the anomalous skin effect. (a) Typical trajectory (b) 'Effective' trajectory.

Figure 3.3 shows a typical trajectory, path (a), between collisions, and along this path E varies from zero deep inside the metal to something approaching its surface value inside the skin-depth region. In this case, $\mathbf{J}(\mathbf{r})$ is governed not simply by $\mathbf{E}(\mathbf{r})$, but rather by the value of $\mathbf{E}$ throughout a volume of radius something like l surrounding the point $\mathbf{r}$. We must therefore replace Ohm's Law by a *non-local relation* of the form

$$\mathbf{J}(\mathbf{r}) = \int K(\mathbf{r} - \mathbf{r}')\mathbf{E}(\mathbf{r}')\, \mathrm{d}^3\mathbf{r}'. \tag{3.11}$$

Here the *kernel* $K(\mathbf{r} - \mathbf{r}')$ must have a *range* of about l. That is, K must drop to zero for distances $|\mathbf{r} - \mathbf{r}'| > l$, as sketched in Fig. 3.4.

It is worth noting that eq. (3.11) reduces to a local form when the range of K is much less than the distance over which $\mathbf{E}$ varies, that is, in the normal skin effect limit $l \ll \delta$. In this case, as sketched in Fig. 3.4, $\mathbf{E}$ has effectively the same value at all points within the range of K. We may therefore, without appreciable error, replace $\mathbf{E}(\mathbf{r}')$ by $\mathbf{E}(\mathbf{r})$ in the integral. $\mathbf{E}$ then comes outside the integral, to give

$$\mathbf{J}(\mathbf{r}) = \sigma\mathbf{E}(\mathbf{r}) \tag{3.12}$$

with

$$\sigma = \int K(\mathbf{r}-\mathbf{r}')\,d^3\mathbf{r}'. \tag{3.13}$$

That is, we have a local relation.

A proper theoretical analysis of the anomalous skin effect was first given by Reuter and Sondheimer (1948). They obtained a non-local relation between $\mathbf{J}$ and $\mathbf{E}$, with the detailed form

$$\mathbf{J}(\mathbf{r}) = \frac{3\sigma}{4\pi l}\int \frac{\mathbf{R}\{\mathbf{R}\cdot\mathbf{E}(\mathbf{r}')\}}{R^4}\,\exp\left(-R/l\right)d^3\mathbf{r}', \tag{3.14}$$

where $\mathbf{R} = \mathbf{r}-\mathbf{r}'$ and $R = |\mathbf{R}|$. Equation (3.14) is of the general form of eq. (3.11), and the kernel does indeed have a range l, as we anticipated.

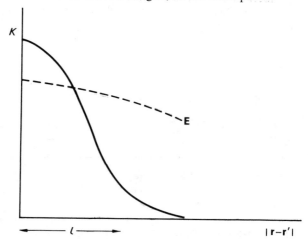

FIG. 3.4   Kernel $K(\mathbf{r}-\mathbf{r}')$ with range l. Dashed line—sketch of variation of E for the normal skin effect limit $\delta \gg l$.

As we have indicated, experiments on the anomalous skin effect are carried out at liquid-helium temperatures, since the mean free path is then limited by scattering off impurities, whereas at room temperature the mean free path is limited by phonon scattering, and is much shorter. In a well-purified specimen, the low-temperature mean free path can be three or four orders of magnitude longer than the room-temperature mean free path. Fuller treatments of the anomalous skin effect than we have space for are given by Kittel (1968) and Pippard (1960), among others. We content ourselves with observing that electrons travelling along paths like (a) in Fig. 3.3 absorb far less energy from the microwave field than electrons travelling along paths like (b), more or less parallel to the surface. Pippard named the former *ineffective*, and the latter *effective trajectories*. The surface impedance therefore depends on the dynamics of electrons travelling in particular directions, so that in a single-crystal specimen the anomalous skin effect is a tool for studying Fermi surface properties. Further details are given in the article by Pippard (1960).

3.3   London equations for a superconductor

We now turn to the problem of setting up a theoretical framework which will account for the Meissner effect. Two basic ideas are needed. First, we need the fundamental property which superconductivity has in common with superfluid flow in He II, that the flow can be characterized by a macroscopic wave function ψ. This led us in Chapter 2 to write the particle current density in He II as

$$\mathbf{J}_P = -\frac{i\hbar}{2m_4}(\psi^*\nabla\psi - \psi\nabla\psi^*) = \frac{\hbar}{m_4}|\psi|^2\nabla S \tag{3.15}$$

where S is the phase of the wave function ψ. In a superconductor, the supercurrent is an electric current, which as we saw is coupled to the magnetic field. We must therefore modify eq. (3.15) to include this coupling; the result of doing so is a relation between $\mathbf{J}$ and the magnetic vector potential $\mathbf{A}$ which is known as London's equation. Note, however, that eq. (3.15) is a local relation, in the sense that $\mathbf{J}_P$ at a point $\mathbf{r}$ is determined by the value of ∇S at the same point $\mathbf{r}$. London's equation, too, is a local relation. The second idea we need, which is due to Pippard, is that a local relation is not adequate to describe a pure superconductor. Just as Ohm's Law has to be replaced by a non-local relation for the anomalous skin effect, so London's equation has to be replaced, for a pure superconductor, by a more general non-local relation. This relation is known as Pippard's equation. As might be expected, the range of the kernel in Pippard's equation becomes the mean free path l in a sufficiently impure alloy; this means of course that for an alloy Pippard's equation can simplify to a local relation, namely London's equation. In broad outline, then, London's equation holds for alloys, and Pippard's for pure superconductors; we shall be able to give a more precise statement of the respective ranges of validity towards the end of the next section.

This section will be devoted to London's equation, with the understanding that we are talking about alloys. We shall deal with the more general Pippard equation in the next section.

We can modify eq. (3.15) to allow for the fact that in a superconductor the current is one of charged particles by making the usual quantum-mechanical replacement

$$\nabla \to \nabla \pm \frac{2ie}{\hbar}\mathbf{A}. \tag{3.16}$$

We take the $+$ sign if ∇ acts on ψ^*, and the $-$ sign if ∇ acts on ψ. Here $\mathbf{A}$ is the vector potential of any magnetic field $\mathbf{B}$ present, for example in the penetration region or the mixed state, defined so that

$$\mathbf{B} = \text{curl}\,\mathbf{A}. \tag{3.17}$$

Equation (3.16) contains the charge $2e$ because the carriers are pairs of electrons; the use of $2e$ for the charge is very well supported by the experiments on quantization of magnetic flux, which we discuss in Chapter 4. With eq. (3.16), the current, eq. (3.15) becomes

$$\mathbf{J}_e = \frac{e\hbar}{m}|\psi|^2\nabla S - \frac{2e^2}{m}|\psi|^2\mathbf{A}. \tag{3.18}$$

We work in terms of the electric current $\mathbf{J}_e = 2e\mathbf{J}_P$, and we have replaced m_4 by $2m$, the mass of a Cooper pair.

Equation (3.18) is the most general form of London's equation. The way in which we use it depends on whether or not we are dealing with a specimen which is multiply connected, that is, has holes in it. The reason is that in a multiply connected specimen, S need not be single valued, whereas in a simply connected specimen (no holes) it must be. To see this, consider a path C round a hole in a multiply connected specimen, as in Fig. 3.5. S is the phase of ψ, and we can increase

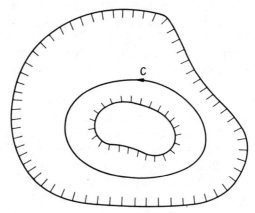

FIG. 3.5   A superconducting specimen with a hole in it, and a path C round the hole.

S by any multiple of 2π without altering ψ. Hence if we go once round C, S need not come back to the same value; we have merely

$$\oint_C \nabla S \cdot d\mathbf{l} = 2n\pi. \tag{3.19}$$

It is easy to see (problem 3.4) that one gets the same value of n for any path C encircling the same hole once, so that n is a characteristic property of the hole. We shall see, in§4.1, that n is the number of quanta $h/2e$ of magnetic flux trapped in the hole. In a simply connected specimen, on the other hand, we must have

$$\oint_C \nabla S \cdot d\mathbf{l} = 0 \tag{3.20}$$

(see problem 3.4), and S is single valued. Multiply connected geometries are rather important in superfluidity and superconductivity, and we shall have a lot to do with them in Chapters 4 and 5. For the time being, however, we restrict our attention to a simply connected specimen, in which S is single valued. In this case, as we shall now see, we can eliminate the ∇S-term from eq. (3.18).

The elimination of the ∇S-term hinges upon a *gauge transformation*. If we put

$$\mathbf{A} = \mathbf{A}_1 + \nabla\chi \tag{3.21}$$

$$S = S_1 + 2e\chi/\hbar \tag{3.22}$$

where χ is a scalar function, then the values of $\mathbf{B}$ and $\mathbf{J}$ are unaltered. Since $\mathbf{B}$ and $\mathbf{J}$ are what we can measure, we can use the transformation to go from a given $\mathbf{A}$ and S to an $\mathbf{A}_1$ and S_1 which we may find more convenient. In particular, for a simply connected superconductor, we can choose a gauge function χ which makes S_1 zero: we just have to take $\chi = \hbar S/2e$, where S is our original phase function. With this choice of gauge, eq. (3.18) becomes

$$\mathbf{J}_e = -\frac{2e^2}{m}|\psi|^2\mathbf{A}_1, \tag{3.23}$$

which is what is usually known as London's equation. In a multiply connected specimen, on the other hand, the use of the gauge function $\chi = \hbar S/2e$ would lead to a vector potential $\mathbf{A}_1$, which was not single valued, so it is easier then to retain the phase S explicitly.

Our task now is to see that eq. (3.23) does lead to the exclusion of flux from a simply connected superconductor. If we take the curl of this equation, we find

$$\mathbf{B} + \frac{m}{2e^2 n_C}\operatorname{curl}\mathbf{J}_e = 0, \tag{3.24}$$

where $n_C = |\psi|^2$ is the density of Cooper pairs. To see that eq. (3.24) does describe the exclusion of magnetic flux, we eliminate curl $\mathbf{J}$ between it and the curl of the Maxwell equation

$$\operatorname{curl}\mathbf{B} = \mu_0\mathbf{J}_e \tag{3.25}$$

to get

$$\nabla^2\mathbf{B} = \mathbf{B}/\lambda_L^2 \tag{3.26}$$

with

$$\lambda_L^2 = m/2e^2\mu_0 n_C. \tag{3.27}$$

For a plane surface in a parallel field the solution is

$$\mathbf{B} = \mathbf{B}_0 \exp(-x/\lambda_L), \tag{3.28}$$

where $\mathbf{B}_0$ is the value at the surface, and x measures distance in from the surface. This is the penetration law we stated in §1.2: the flux density is zero inside the superconductor, and decays from its external value in the penetration region. λ_L appears here as the penetration depth.

The crucial assumption we made in arriving at London's equation, and consequently deriving the Meissner effect, is that the superconductor is characterized by a macroscopic wave function. This is exactly the line of reasoning that led London to postulate the equation in the first place. Equation (3.23) describes the macroscopic quantum-mechanical current, and the Meissner effect is just the exclusion of magnetic flux by the quantum current.

We can rewrite eq. (3.27) for the penetration depth in terms of the density of superconducting electrons $n_s = 2n_C$:

$$\lambda_L^2 = m/e^2\mu_0 n_s \qquad (3.29)$$

which links the temperature dependence of λ_L to that of n_s. The Gorter–Casimir theory gave $n_s \propto 1-t^4$, which leads to $\lambda \propto (1-t^4)^{-1/2}$, quoted in § 1.2. This gives an excellent first-order description of experimental results, as shown in Fig. 3.6.

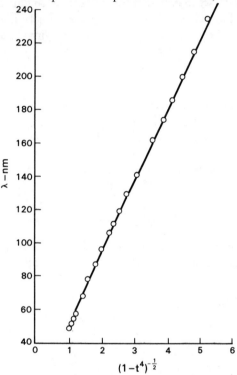

FIG. 3.6   Penetration depth of Pb. (After Gasparovic and McLean, 1970.)

There are two principal experimental techniques for measurement of the penetration depth. First, there is the inductance technique, which we mentioned in § 1.2. Secondly, the penetration depth can be obtained from microwave surface-impedance measurements. The simplest view of the microwave impedance of a superconductor would be that during the a.c. cycle the magnetic flux moves in and out of the penetration depth region, so that the impedance is purely inductive, with a reactive skin depth equal to the penetration depth λ. This view is essentially correct, but two minor corrections are needed: at microwave frequencies the quasiparticles, or normal-fluid component, can be accelerated, so that there is some resistance, and in addition the reactive skin depth δ_i can differ by something like 10% from the zero frequency value λ. The surface-impedance method is discussed very fully by Waldram (1964); in particular he uses the Kramers–Kronig relations to derive λ from δ_i and the values of the surface resistance over a range of frequencies.

We also mentioned in § 1.2 that in alloys, λ increases with increasing impurity content. An example is shown in Fig. 3.7. This increase in λ as the mean free path l

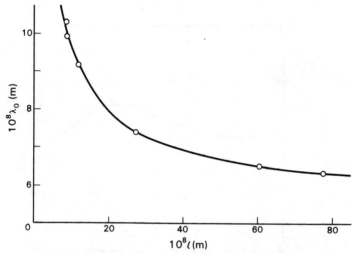

FIG. 3.7 Variation of zero-temperature penetration depth λ_0 with mean free path in Sn with In impurity. (After Pippard, 1953.)

decreases implies that the value of n_s one uses in eq. (3.29) should decrease as l decreases. From the point of view of the microscopic theory, one can understand this as follows. To exclude the flux, the superelectrons must move as a screening current; in particular, they must move all with one momentum **p**. This is of no consequence in the pure material, where the condensation is into states of given momentum. However, in an alloy, the condensation is into states ϕ_n, which do not have an exact momentum (see Appendix I). This means that the proportion of the condensate contributing to the screening current decreases as l decreases. The situation is somewhat analogous to that in He II, where interparticle interactions deplete the condensate, although here the condensate is not depleted but ineffectual.

3.4 Pippard's equation

The London equation, which we have discussed in the previous section, was used to describe the electrodynamics of superconductors until about 1950. At that time, Pippard started a series of measurements of the microwave surface impedance of superconductors; his most important conclusion, as we have already mentioned, was that for pure superconductors the London equation should be replaced by a non-local equation.

Perhaps the simplest evidence that the London equation is inadequate comes from Pippard's measurement of the anisotropy of the penetration depth in pure Sn. Superconducting Sn is an orthorhombic structure, that is, the unit cell has the form of a 'sat-on cube', with two axes (the a-axes) equal and one (the c-axis) of a different length. Figure 3.8 shows the penetration depth as a function of angle in

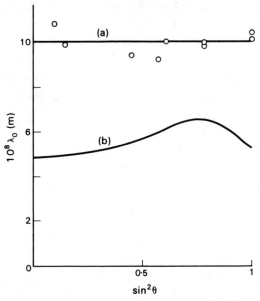

FIG. 3.8   Anisotropy of zero-temperature penetration depth in Sn. (a) Sn + 3% In, (b) pure Sn, θ is the angle between the c-axis of the crystal and the supercurrents flowing in the penetration depth region. (After Pippard, 1953.)

both a pure and an alloy specimen. The significant point is that λ varies with angle, and is not symmetric about the c-axis, in the pure specimen. In the impure specimen, there is apparently no angular variation of λ. One can allow for some anisotropy, within a local theory, if one replaces λ_L^{-2} by a second-rank tensor, so that eq. (3.25) becomes

$$\nabla^2 B_i = \sum_{j=1}^{3} T_{ij} B_j. \tag{3.30}$$

However, the tensor T_{ij} must be symmetric about the c-axis (see the book by Nye (1957), for example), and the penetration depth calculated from eq. (3.30) is also symmetric about the c-axis. It is obvious, therefore, that the anisotropy of λ shown for the pure specimen in Fig. 3.8 cannot be described by this means.

The anisotropy of λ is sufficient reason on its own for going beyond a local equation of the London type. Pippard argued further that to give the right surface energy there must in any case be a range of coherence ξ to the superconducting state. In Chapter 4 we shall consider the surface energy of an interface between the normal and superconducting phases; we shall see there that if the superconducting phase disappears abruptly at the interface the surface energy is negative, and of order $\mu_0 H_{cb}^2 \lambda/2$. In a pure superconductor, the surface energy is in fact positive; this can only be the case if the boundary of the superconducting phase is spread over a length ξ. The surface energy is then $\mu_0 H_{cb}^2 (\xi - \lambda)/2$, which is positive as required for $\xi > \lambda$. Pippard took the view that this same length should also come into the electromagnetic response, so that $\mathbf{J}(\mathbf{r})$ would be determined by the values of $\mathbf{A}$ within a distance ξ of the point $\mathbf{r}$. We shall emphasize later that it is important in general to distinguish between the coherence length for the electromagnetic response and the coherence length for the surface energy, because they are not always the same. For the time being, however, we simply need the idea that the response of $\mathbf{J}$ to $\mathbf{A}$ should be non-local, with a kernel of some range ξ.

The relation Pippard proposed to replace the London equation is

$$\mathbf{J}(\mathbf{r}) = -\frac{3ne^2}{4\pi\xi_0 m} \int \frac{\mathbf{R}\{\mathbf{R}\cdot\mathbf{A}(\mathbf{r}')\}}{R^4} \exp\left(-R/\xi_P\right) d^3 r', \tag{3.31}$$

where $\mathbf{R} = \mathbf{r} - \mathbf{r}'$ and $R = |\mathbf{R}|$. The coherence length ξ_P is given by

$$\frac{1}{\xi_P} = \frac{1}{\xi_0} + \frac{1}{l}, \tag{3.32}$$

where ξ_0 is a constant of the pure material; in particular $\xi_P = \xi_0$ in the pure material, and $\xi_P = l$ for $l \ll \xi_0$. The detailed form was of course based on an analogy with eq. (3.14), the Reuter–Sondheimer equation for the anomalous skin effect; however, the later microscopic theory produced a relation essentially identical to Pippard's equation. There is one important difference between Pippard's equation and the Reuter–Sondheimer equation. In both cases the kernel has the range l for an alloy, but in the Pippard equation the range has the finite value ξ_0 for the pure material. The microscopic theory gives

$$\xi_0 = 0.18\hbar v_F/k_B T_c \tag{3.33}$$

where v_F is the Fermi velocity. This value is not a surprising one, in the sense that $k_B T_c$ is the fundamental energy characterizing the superconducting state, so that $\hbar/k_B T_c$ is a natural time interval to occur.

We can now see when the London equation is valid, and when we should use the Pippard equation. As we saw in Fig. 3.4, eq. (3.31) will reduce to a local relation between $\mathbf{J}$ and $\mathbf{A}$, when ξ_P is small compared to the distance over which $\mathbf{A}$ varies, namely λ. That is, a London equation holds whenever

$$\lambda(T) \gg \xi_P \tag{3.34}$$

where we show the temperature argument to stress that λ is temperature dependent, while ξ_P is not. Since $\lambda \to \infty$ as $T \to T_c$, there is a London temperature interval near to T_c even in a pure superconductor. The more important result is that as l decreases, ξ_P decreases and $\lambda(T)$ increases, so that eq. (3.34) holds at all temperatures in a sufficiently impure alloy. In the limit $l \ll \xi_0$, we have $\xi_P = l$; with $\mathbf{A}$ taken outside the integral, eq. (3.31) becomes

$$\mathbf{J}(\mathbf{r}) = -\frac{ne^2}{m}\frac{l}{\xi_0} \mathbf{A}(\mathbf{r}), \tag{3.35}$$

which correctly gives the increase of λ as l decreases.

3.5   Superconductors in microwave and infrared fields

The previous two sections dealt with the response of a superconductor to a d.c. magnetic field, in particular the exclusion of the flux by the dissipation-free quantum current. We now turn to the behaviour in a.c. fields, that is, the response to microwave and infrared radiation. In contrast to the d.c. case, the a.c. response is dissipative, and the experimental results we shall discuss consist of measurements of absorption as a function of frequency and temperature.

The mechanism for absorption of radiation of frequency ω at temperature T depends on the ratio $\hbar\omega/2\Delta\,(T)$. For $\hbar\omega < 2\Delta\,(T)$, energy can be absorbed by the acceleration of the thermally excited quasiparticles, or normal fluid component. This is the mechanism which we discussed towards the end of §3.3 as giving a small resistive component to the surface impedance. For $\hbar\omega > 2\Delta\,(T)$ absorption of energy by the normal fluid still occurs. In addition, however, it is possible for a photon with energy greater than $2\Delta\,(T)$ to break up a Cooper pair and create two quasiparticles. This additional mechanism leads to a strong increase in the absorption as ω increases past $2\Delta\,(T)/\hbar$.

The value of the threshold frequency $\omega_{\mathrm{g}} = 2\Delta\,(T)/\hbar$ for creation of quasiparticles depends, of course, on the temperature. At absolute zero, $\omega_{\mathrm{g}} = 3 \cdot 5 k_{\mathrm{B}}\,T_{\mathrm{c}}/\hbar$, which for a critical temperature of a few degrees, gives a threshold in the far infrared. As T increases, Δ decreases (Fig. 1.25), approaching zero at T_{c}. Thus ω_{g} moves down, eventually into the microwave region, as T increases. The general behaviour is well illustrated by Fig. 3.9, which shows the microwave surface resistance of Zn as a function of temperature at two different frequencies. At the

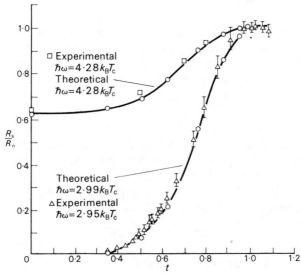

FIG. 3.9   Microwave surface resistance R_{s} of Zn at two frequencies (R_{n} = normal state value). Theoretical curves from microscopic theory. (After Zemon and Boorse, 1966, with theoretical curves calculated from the work of Miller, 1960.)

higher frequency, $\hbar\omega > 2\Delta\,(T = 0)$, and consequently there is absorption even at $T = 0$. At the lower frequency, $\hbar\omega < 2\Delta\,(T = 0)$, and there is no absorption at $T = 0$. As T increases at the lower frequency, the absorption increases, slowly at first when absorption is only by excited quasiparticles, and then more rapidly as the threshold $\hbar\omega = 2\Delta\,(T)$ is passed. At both frequencies, of course, the surface resistance reaches the normal-state value at $T = T_c$. We should mention finally that microscopic calculations, the most recent and most general being by Wong (1967), agree very well with the experimental results, as can also be seen in Fig. 3.9.

A typical infrared absorption curve is shown in Fig. 3.10, showing the expected sharp increase in absorption at $\omega_g = 2\Delta/\hbar$. The experiments are difficult to

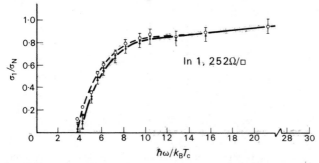

FIG. 3.10   Infrared absorption of In as a function of frequency. Vertical axis: real (absorptive) part of conductance, divided by normal state value. (After Ginsberg and Tinkham, 1960.)

perform, because the threshold is generally in the far infrared; historically, however, infrared absorption was important, because it gave the first direct evidence for the existence of the energy gap.

3.6   Electron tunnelling

We have seen that the infrared absorption, and to some extent the microwave absorption, are governed by the energy gap. It was first shown by Giaever (1960a, b) that tunnelling through an insulating barrier from one superconductor to another, or to a normal metal, gives a very direct measurement of the energy gap.

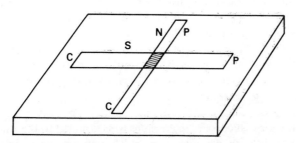

FIG. 3.11   Typical NS junction, evaporated on a glass substrate. The junction occupies the shaded region. N—normal metal, S—superconducting metal. Current leads are attached to CC; potential leads to PP.

The usual experimental set up is sketched in Fig. 3.11 for a normal-superconductor (NS) junction. First a normal metal strip, typically Al or Mg, is evaporated on to a substrate, then the surface is oxidized in air or oxygen to make the insulator, and finally the superconducting strip is evaporated. Current and potential leads are attached as shown.

We first consider tunnelling between two normal metals. On a potential-energy

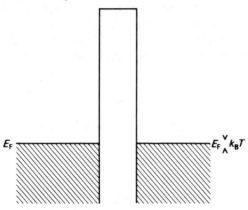

FIG. 3.12   Potential-energy diagram for NN tunnelling. The barrier height is typically about $10^4 k_B T$ at helium temperature.

diagram, Fig. 3.12, the insulator presents a barrier, typically 10 eV or so high, to the passage of electrons. Since the electrons are confined to a region around $k_B T$ from the Fermi surface, they cannot pass over the barrier. However, if a voltage is applied between the metals, some current flows by quantum-mechanical tunnelling through the barrier. The current is proportional to the voltage (for modest voltages). Tunnelling depends on overlap in the barrier region of the electronic wave functions appropriate to either side; since the wave functions decay exponentially in the barrier, the tunnelling resistance increases exponentially with barrier thickness.

Tunnelling between normal metals does not give much information about the electronic states at the Fermi surface. The tunnelling probability W is proportional to the density of states $N(0)$ at the Fermi surface, and to the average velocity $\bar{v}$ towards the barrier. The latter is proportional to v_F, the Fermi velocity, so the tunnelling current is proportional to $N(0)v_F$. However, it is well known in Fermi surface theory (see Pippard, 1960) that $N(0)v_F$ is simply proportional to the total area of the Fermi surface. Since the current is also exponentially dependent on the unmeasurable barrier thickness, no Fermi surface properties can be deduced from the tunnelling resistance.

The situation is quite different in superconductors, and in fact tunnelling has been one of the most successful techniques to be applied to the study of superconductivity. To begin with, we must distinguish between two main types of tunnelling. First, if there is a very thin barrier between two superconductors, the condensate wave functions of the two sides overlap, and in consequence a weak supercurrent can flow between the two sides. This is known as Josephson tunnelling, after its discoverer; we shall not go into Josephson tunnelling any further at the moment, since Chapter 5 is devoted to the subject. The kind of barrier that is typically used in Josephson tunnelling is the very thin oxide that forms on the surface of tin; thus if a tin film is evaporated, then allowed to oxidize, and a second superconducting film is evaporated as in Fig. 3.11, the resulting junction may be expected to carry a weak supercurrent. The oxide thickness is often quoted as 1 nm.

The second type of tunnelling, which concerns us now, is the tunnelling of quasiparticles through the barrier. We recall that quasiparticles are created when Cooper pairs break up; we have previously referred to the thermally excited quasiparticles as the normal-fluid component. Quasiparticle tunnelling is studied with rather thicker oxide barriers than are used for Josephson tunnelling; a typical barrier is the oxide that forms on the surface of freshly evaporated Al or Mg. Quasiparticle tunnelling may be studied either between two superconductors (SS junction), or between a superconductor and a normal metal (SN junction), whereas Josephson tunnelling, of course, occurs only between two superconductors.

In general, quasiparticle tunnelling reflects the density of states in the superconductor or superconductors involved. Since it takes energy 2Δ to create two quasiparticles from a Cooper pair, the density of quasiparticle states in a superconductor is zero for a distance Δ above and below the Fermi surface. The simplest form of the BCS theory gives for the density of states $N(E)$

$$N(E) = 0 \qquad\qquad |E| < \Delta$$

$$= \frac{N(0)}{(E^2 - \Delta^2)^{1/2}} \qquad |E| > \Delta, \qquad\qquad (3.36)$$

where the energy E is measured from the Fermi surface, and $N(0)$ is the density of states at the Fermi surface in the normal metal. $N(E)$ is sketched in Fig. 3.13; the states which in the normal metal are within a distance Δ of the Fermi surface have been moved out and piled up outside the energy-gap region.

With such a radical modification of the density of states, the tunnelling characteristic is naturally quite different from the normal-state one. Fig. 3.14 shows the characteristics at various temperatures for a symmetric SS junction. At $T = 0$ the current is (ideally) zero up to the voltage V_0 given by $eV_0 = 2\Delta$, typically of the order of millivolts. The current then increases rapidly, finally approaching the Ohmic line appropriate to the normal state. For $T > 0$, there is some current for $V < V_0$, and still a rapid increase at $V = V_0$. These characteristics can be understood in terms of the level differences between the density of states on either side — see Fig. 3.15. A voltage difference V between the two sides of the junction corresponds to a relative displacement of the Fermi levels by eV. At $T = 0$ all the states below E_F (shaded in Fig. 3.15) are occupied, and all the states above E_F are empty. There is therefore no current until $V_0 = 2\Delta/e$, at which point the current increases rapidly since a high density of occupied states then lies opposite a high density of unoccupied states. For $T > 0$ there is some thermal occupation of the

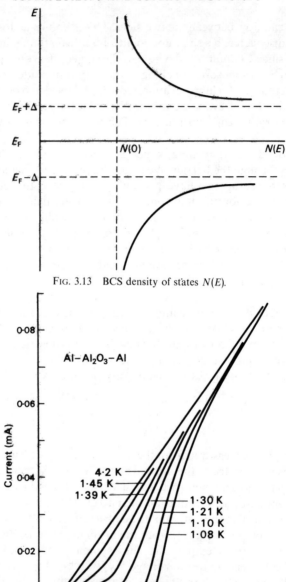

FIG. 3.13   BCS density of states $N(E)$.

FIG. 3.14   Current–voltage characteristics of an Al–Al junction at various temperatures. (After Giaever and Megerle, 1961.)

states above $E_F + \Delta$, and some depletion of the states below $E_F - \Delta$, as sketched in Fig. 3.16. Consequently, some current flows even for biases $V < 2\Delta/e$.

Similar arguments explain the Al–Pb characteristics shown in Fig. 3.17. At high

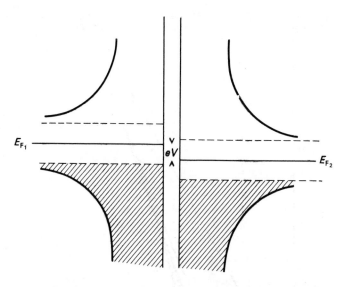

FIG. 3.15 Energy-level diagram for SS tunnelling at $T = 0$.

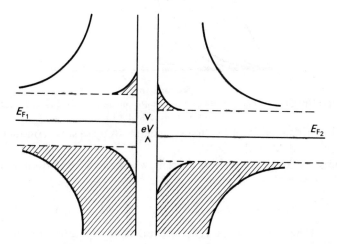

FIG. 3.16 Energy-level diagram for SS tunnelling at $T > 0$.

temperatures (above 1·2 K) the Al is normal, and we have SN tunnelling, whilst at the lowest temperature (1·05 K), the Al is superconducting, so we have an SS′ (two different superconductors) characteristic. In the SN junction, the current increases sharply at $eV = \Delta$, since it is here that occupied states in the superconductor come opposite unoccupied states in the normal metal. In the SS′ junction, there is in principle a maximum at $eV = \Delta_1 - \Delta_2$ and a minimum at $eV = \Delta_1 + \Delta_2$; the region in between includes a part with a negative differential resistance, which does not show up with the circuitry used to obtain the results of Fig. 3.17. We note that Figs.

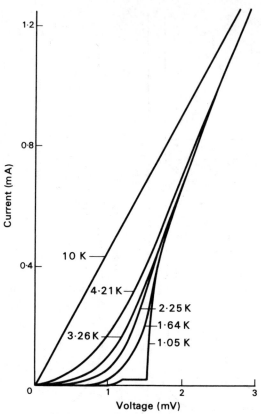

FIG. 3.17   Current–voltage characteristics of an Al–Pb junction at various temperatures. (After Giaever and Megerle, 1961.)

3.14 and 3.17 are taken from the first systematic study of superconductive tunnelling (Giaever and Megerle, 1961).

An expression for the tunnelling current, which includes all the above examples as special cases, is

$$I(V) = I_0 \int N_1(E + eV) N_2(E) \{ f(E) - f(E + eV) \} \, dE \tag{3.37}$$

(Cohen et al., 1962), where N_1 and N_2 are the densities of states on either side, and f is the Fermi function. We may regard diagrams like Figs. 3.15 and 3.16 as illustrative of eq. (3.37).

Tunnelling has been used extensively, particularly by Rowell and co-workers, to study deviations from the simple density of states given in eq. (3.36). That density of states is derived from the BCS theory in its simplest form, which involves replacing a complicated phonon-mediated interaction by a constant. If measurements are made, not of I versus V, but of dI/dV or dV/dI versus V, then tunnelling is a very sensitive probe of the details of the actual phonon interaction. The derivative, dV/dI for example, can be obtained by adding a small a.c. current to the d.c. current I, and using a phase-sensitive technique to detect the a.c. voltage produced. The second derivative, d^2V/dI^2, is given by the signal generated at twice the frequency used for the a.c. current. Some idea of the amount of structure that can be seen in such measurements can be obtained from Figs. 3.18 and 3.19 (taken from Clark, 1968), which show first and second derivative curves for Mg–Tl

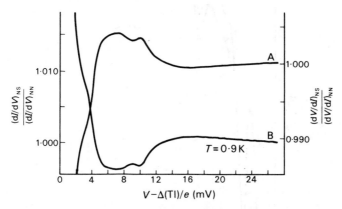

FIG. 3.18   Relative conductance and resistance of an Mg–Tl junction as a function of applied voltage V. A: dV/dI, B: dI/dV. (After Clark, 1968.)

junctions. All the structure in such measurements can be interpreted in terms of properties of the phonons.

For the case of an NS junction at $T \ll T_c$ in particular, eq. (3.37) gives a simple result for dI/dV. The derivative involves $f'(E + eV)$, which is very sharply peaked at $E + eV = E_F$. Since $N_1(E)$ is constant for the normal metal, we get simply

$$dI/dV \propto N_2(E_F - eV). \tag{3.38}$$

The derivative measures the density of states directly. The effect on $N(E)$ of retaining the full details of the phonon interaction has been worked out by Scalapino and McMillan, using a formulation of the problem due to Eliashberg. The deviations from the BCS form are strongest in the *strong-coupling* superconductors like Pb, Tl and Hg in which the ratio $k_B T_c/\hbar\omega_D$ is relatively large. The form of $N(E)$ is governed by the phonon density of states $n_p(E)$, and the forms obtained from superconductive tunnelling for $n_p(E)$ are in good agreement with the results of inelastic neutron scattering.

The investigation of strong-coupling superconductivity by tunnelling has been the most thorough and detailed test of the microscopic theory. Full details, both theoretical and experimental, are given in the articles by Scalapino and by Rowell and McMillan in Parks (1969). The book by Solymar (1972) contains an extended account of both quasiparticle and Josephson tunnelling.

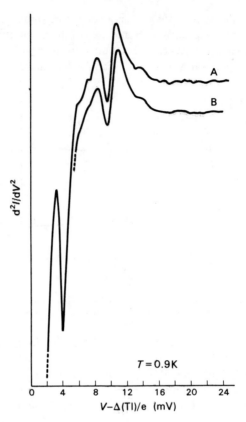

FIG. 3.19 d^2I/dV^2 against V for an Mg–Tl junction. A: 0·9 mV peak to peak a.c. signal; B: 1·5 mV peak to peak a.c. signal. (After Clark, 1968.)

Problems

3.1   Derive eq. (3.4) from the three previous equations.

3.2   Confirm that the inequality $\sigma \gg \omega\varepsilon$ is dimensionally correct, and check from published values that the inequality holds for Cu and Al at room temperature (take $\varepsilon = \varepsilon_0$ in both cases).

3.3   Estimate the normal skin depth δ for a frequency $\omega = 2\pi \times 10^{10}\,\text{s}^{-1}$ (3 cm radiation) at helium temperature for a Cu specimen with a resistance ratio $R_{300\,\text{K}}/R_{4\cdot2\,\text{K}}$ of 10^4. Find also the low-temperature mean free path l, and confirm that $\delta \ll l$.

(Room-temperature resistivity of Cu $= 1\cdot72 \times 10^{-8}\,\Omega\text{m}$; conversion from resistivity to mean free path for Cu is $\rho l = 1\cdot36 \times 10^{-15}\,\Omega\text{m}^2$).

3.4   Show that

$$\oint_C \nabla S \cdot d\mathbf{l} = 2n\pi$$

holds, with the same value of the integer n, for *any* contour C encircling the hole in Fig. 3.5. (Note that any two such contours C_1 and C_2 can be deformed continuously into one another). Show likewise that

$$\oint_C \nabla S \cdot d\mathbf{l}$$

must be zero for any contour C which does not encircle a hole. (Shrink the contour continuously to zero).

3.5 Confirm that the gauge transformation of eqs. (3.21) and (3.22) does leave **B** and **J** unaltered.

3.6 Derive eq. (3.26) and show that eq. (3.28) is the solution for a plane surface.

3.7 Consider a loop of wire carrying a supercurrent. Write London's equation in integral form:

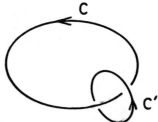

$$\int \mathbf{H} \cdot d\mathbf{S} = -\lambda_L^2 \oint_C \mathbf{J} \cdot d\mathbf{l}, \quad (1)$$

where the contour C is the wire, and write the appropriate Maxwell equation in integral form

$$\int \mathbf{J} \cdot d\mathbf{S} = \oint_{C'} \mathbf{H} \cdot d\mathbf{l}, \tag{2}$$

where C' is any contour encircling the wire. Show that the implication of the equations is that **H** 'generates' **J** in the same way that **J** always generates **H**. This gives a way of seeing that persistent supercurrents are implied by London's equation.

3.8 Show that eq. (3.35) follows from eq. (3.31) in the local limit $(\lambda(T) \gg \xi_P)$ when $\xi_P = l$. (Take **J** and **A** both in the z-direction).

3.9 Mark on Fig. 3.9 the temperature at which $2\Delta(T) = 3 \cdot 0 k_B T_c$ (use Fig. 1.22) and confirm that this coincides with the sharp increase in absorption for $\hbar\omega = 2 \cdot 95 k_B T_c$.

3.10 Draw density of states diagrams like Figs. 3.15 and 3.16 for the SN junction, and confirm that the current increases sharply at $eV = \Delta$.

3.11 Draw density of states diagrams for SS' tunnelling. Confirm in particular that there is a maximum in the current at $eV = \Delta_1 - \Delta_2$ and a minimum at $eV = \Delta_1 + \Delta_2$.

3.12 Show that eq. (3.37) is equivalent to Figs. 3.15 and 3.16 for SS tunnelling.

References

CLARK, T. D., 1968, *J. Phys.* C **1**, 732.
COHEN, M. H., FALICOV, L. M. and PHILLIPS, J. C., 1962, *Phys. Rev. Lett.* **8**, 316.
DHEER, P. N., 1961, *Proc. Roy. Soc.* **A260**, 333.

GASPAROVIC, R. F. and MCLEAN, W. L., 1970, *Phys. Rev.* **B2**, 2519.

GIAEVER, I., 1960a, *Phys. Rev. Lett.* **5**, 147.

GIAEVER, I., 1960b, *Phys. Rev. Lett.* **5**, 464.

GIAEVER, I. and MEGERLE, K., 1961, *Phys. Rev.* **122**, 1101.

GINSBERG, D. M. and TINKHAM, M., 1960, *Phys. Rev.* **118**, 990.

GUYON, E., 1966, *Adv. Phys.* **15**, 417.

KITTEL, C., 1968, *Introduction to Solid State Physics* (Wiley, New York.).

MILLER, P. B., 1960, *Phys. Rev.* **118**, 928.

NYE, J. F., 1957, *Physical Properties of Crystals* (Oxford University Press, London).

PARKS, R. D., 1969, *Superconductivity*, 2 vols. (Dekker, New York).

PIPPARD, A. B., 1953, *Proc. Roy. Soc.* **A216**, 547.

PIPPARD, A. B., 1960, *Reports on Progress in Physics*, **23**, 176.

REUTER, G. E. H. and SONDHEIMER, E. H., 1948, *Proc. Roy. Soc.* **A195**, 336.

SOLYMAR, L., 1972, *Superconductive Tunnelling and Applications* (Chapman and Hall, London).

WALDRAM, J. R., 1964, *Adv. Phys.* **13**, 1.

WONG, M. K. F., 1967, *J. Math. Phys.* **8**, 1443.

ZEMON, S. A. and BOORSE, H. A., 1966, *Phys. Rev.* **146**, 309.

Chapter 4

Vortex States

4.1 Quantization of circulation

We have seen already, in Chapters 2 and 3, that the description of a superfluid in terms of a single wave function, $|\psi| \exp(iS)$, leads us in a straightforward way to write the superfluid currents as

$$\mathbf{j_s} = \hbar |\psi|^2 \nabla S \qquad \text{(He II)} \quad (4.1)$$

$$\mathbf{J_e} = \frac{e\hbar}{m}|\psi|^2 \nabla S - \frac{2e^2}{m}|\psi|^2 \mathbf{A} \qquad \text{(superconductor)} \quad (4.2)$$

We use the mass current density $\mathbf{j_s}$ for He II, and the electric current density $\mathbf{J_e}$ for the superconductor. Equation (4.2) implies the normalization $|\psi|^2 = n_s/2$, the density of Cooper pairs. Equations (4.1) and (4.2) hold provided that the flow velocity is small enough for the superfluid density to remain unaltered. In addition, eq. (4.2) is valid only for superconductors of the London type and not for those of the Pippard type.

In Chapter 3 we restricted our attention to singly connected superconductors, and showed that we could use a gauge transformation to eliminate the term in eq. (4.2) involving ∇S. We pointed out at the same time that we could not make this transformation in a multiply connected specimen, one with holes in it. We now study multiply connected regions in both He II and superconductors.

Consider He II occupying an annular region such as the space between two concentric cylinders, as shown in Fig. 4.1. We assume that the temperature is

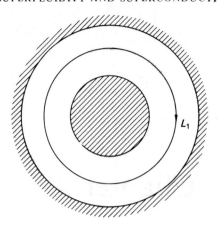

Fɪɢ. 4.1   Superfluid occupying an annular (multiply connected) region; contour L_1 is drawn wholly in the superfluid.

absolute zero, so that the He II is pure superfluid. To find the flow pattern, we examine the quantity known as the circulation

$$\kappa = \oint \mathbf{v_s} \cdot \mathbf{dl}, \tag{4.3}$$

where the integral is taken round any contour wholly within the liquid. It has already been pointed out in Chapter 2 that eq. (4.1) implies that the superfluid velocity can be written

$$\mathbf{v_s} = \frac{\hbar}{m_4} \nabla S, \tag{4.4}$$

and therefore it is possible to express the circulation in terms of the wave-function phase S,

$$\kappa = \frac{\hbar}{m_4} \oint \nabla S \cdot \mathbf{dl}. \tag{4.5}$$

For the circle L_1 (Fig. 4.1) the circulation is

$$\kappa = \frac{\hbar}{m_4} (\Delta S)_{L_1}. \tag{4.6}$$

Since the superfluid wave-function is single-valued, a trip round a closed contour must leave it unchanged, with the result that the change in S can be only an integral multiple of 2π or zero. From eq. (4.6) we see that the circulation is quantized, taking the values

$$\kappa = n \frac{\hbar}{m_4}, \qquad \text{where } n = 0, 1, 2, \dots. \tag{4.7}$$

For obvious reasons h/m_4 is known as the quantum of circulation, and it has the value $9 \cdot 98 \times 10^{-8} \, \mathrm{m^2 \, s^{-1}}$. We note that eq. (4.7) is true for any contour which may be continuously deformed into L_1 without passing outside the boundaries of the liquid, that is for any contour which represents a single trip round the inner cylinder.

The annulus (Fig. 4.1) is an example of a multiply connected region, since it contains what may be regarded as a 'hole' in the superfluid. The presence of holes, that is regions which the superfluid cannot penetrate yet which are completely surrounded by superfluid, ensures that contours such as L_1 in Fig. 4.1 can be drawn. As we have seen, this is a sufficient condition for the quantization of circulation. We now distinguish between two different situations in which this quantization occurs.

Firstly, the holes can be provided by solid boundaries; most commonly the superfluid is enclosed in an annular container (e.g., Fig. 4.39). Arrangements of this kind furnish the most convenient method of studying persistent currents in He II.

Secondly, in rotating He II a hole in the form of a cylinder can appear spontaneously in the superfluid; stable currents are then set up in circles round the cylinder. In this case what we have described is a vortex line (Fig. 4.2). Vortices

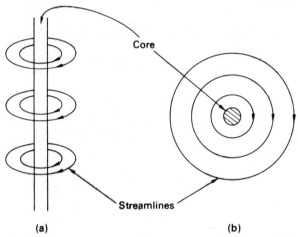

(a)   (b)

FIG. 4.2   (a) Vortex line in He II; (b) transverse section of vortex line.

occur in ordinary liquids and it is not surprising that they are a feature of the superfluid. Indeed, as we shall show, vortices are very easily formed in He II and are of importance in understanding many of its properties.

We now turn to the d.c. magnetic properties of multiply connected superconductors. As in He II, it is convenient to define a *superfluid velocity* $\mathbf{v}_s$ by

$$\mathbf{J}_e = 2e|\psi|^2 \mathbf{v}_s. \tag{4.8}$$

We then rearrange eq. (4.2) as

$$\hbar \nabla S = 2m\mathbf{v}_s + 2e\mathbf{A} \tag{4.9}$$

and integrate it round a contour L_1 encircling a hole in the superconductor (Fig. 4.1). The left-hand side gives

$$\oint_{L_1} \hbar \nabla S \cdot d\mathbf{l}, \quad \text{or} \quad \hbar(\Delta S)_{L_1},$$

where $(\Delta S)_{L_1}$ is the change of phase on going once round L_1. Again we invoke the fact that the wave function is single-valued, so that the phase change is $2n\pi$, where the integer n takes the same value for L_1 or any contour that goes once round the hole. We therefore have

$$\oint_{L_1} 2m\mathbf{v}_s \cdot d\mathbf{l} + \oint_{L_1} 2e\mathbf{A} \cdot d\mathbf{l} = nh \qquad (4.10)$$

as our basic quantization condition, with n characteristic of the hole that L_1 encircles.

We can usually simplify eq. (4.10). If the specimen is a bulk type I superconductor, we can draw the contour L_1 at a distance greater than λ from the hole, so that there is no magnetic field at any point of L_1, and the screening current $\mathbf{J}_s$ is zero. This means that $\mathbf{v}_s$ is zero round L_1, so we have

$$\oint_{L_1} \mathbf{A} \cdot d\mathbf{l} = n \frac{h}{2e}. \qquad (4.11)$$

We transform this line integral to a surface integral by Stokes' theorem

$$\oint_{L_1} \mathbf{A} \cdot d\mathbf{l} = \int_{\Sigma} \operatorname{curl} \mathbf{A} \cdot d\Sigma = \int_{\Sigma} \mathbf{B} \cdot d\Sigma. \qquad (4.12)$$

where the surface Σ spans the contour L_1. The last integral is equal to the total magnetic flux Φ through the hole together with the region surrounding the hole where the field penetrates the superconductor. We therefore have as our final result

$$\Phi = n\phi_0 \qquad (4.13)$$

with

$$\phi_0 = h/2e. \qquad (4.14)$$

The value of ϕ_0, the flux quantum, is $2.07 \times 10^{-15}\,\mathrm{V\,s} = 2.07 \times 10^{-15}\,\mathrm{Wb} = 2.07 \times 10^{-7}\,\mathrm{G\,cm^2}$.

In superconductors, as in He II, the holes round which quantization occurs can be either physical boundaries, or the cores of quantized vortex lines. Vortex lines are not found in all superconductors; as we pointed out in § 1.2, the mixed state of a type II superconductor is a vortex state, while vortices never appear in type I superconductors. There is a close similarity between vortices in He II and superconductors which will become clear as the chapter progresses: in both cases they consist of particle currents circulating round a cylindrical core. On the other hand, there are important distinctions to be drawn, the most obvious being that vortices in superconductors are coupled to the applied magnetic field, whereas a magnetic field has no influence on He II vortices.

The first experiment to demonstrate that circulation is quantized in He II was

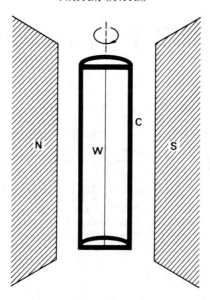

FiG. 4.3   Vinen's apparatus for measuring circulation of He II. C: cylindrical container, W: vibrating
wire, N, S: pole pieces of magnet. (After Vinen, 1961.)

that of Vinen (1961). The apparatus consisted of a cylindrical container with a fine
conducting wire stretched along its axis (Fig. 4.3). The He II filling the container
thus occupied a multiply connected region. Vinen's method exploits the fact that a
solid cylinder, round which there is a fluid circulation, experiences a force when it
moves through the fluid. To see in a simple way how this force arises, we use a
reference frame in which the cylinder is at rest, and an ideal fluid flows past the
cylinder with its attendant circulation (Fig. 4.4). At the point A the circulation
opposes the current, while at B the two flows reinforce one another. Applying
Bernoulli's theorem, we see that the fluid pressure is smaller at B than at A. Taking
account of other points within the circulation, there is a net force on the cylinder in
the direction AB, known as the Magnus force. If we treat the superfluid as an ideal
Euler fluid, the Magnus force on the wire in Vinen's experiment is given by

$$\mathbf{f_M} = \rho_s \boldsymbol{\kappa} \times \mathbf{V}, \tag{4.15}$$

where $\mathbf{V}$ is the velocity of the wire relative to the superfluid outside the circulation,
and $\boldsymbol{\kappa}$ is the vector which indicates the strength of the circulation. Thus $\mathbf{f_M}$ is
transverse to the wire and to its direction of motion. For a fuller treatment of the
Magnus force the reader is referred to a text on hydrodynamics (e.g. Rutherford,
1959).

The apparatus was placed in a magnetic field so that transverse vibrations of the
wire could be excited by the passage of an a.c. electric current through it. When the
surrounding liquid is not rotating, the normal modes of the wire are two waves
plane polarized at right angles to each other and having the same frequency. When
a stable circulation of the superfluid is set up round the wire, the latter is acted

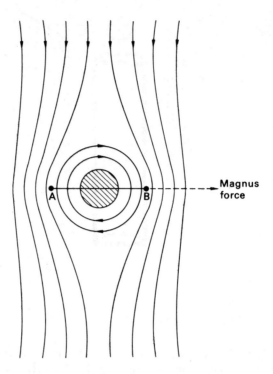

FIG. 4.4   Origin of the Magnus force in an ideal fluid: a solid cylinder has a local circulation round it, and the fluid flows past as indicated by the streamlines. Magnus force acts on the cylinder in the direction shown.

upon by the Magnus force, which causes both planes of vibration to precess. As a result the normal modes of the wire can now be regarded as circularly polarized in opposite senses, with frequencies differing by $\Delta v = \rho_s \kappa / 2\pi W$, where W is the sum of the mass per unit length of the wire plus half the mass of the fluid displaced by this length. The difference Δv appears as a beat frequency of the voltage induced in the wire, and this provides a direct method of measuring the circulation κ.

The superfluid circulation was set up by starting with He I, rotating the container at a steady speed, and then slowly cooling through the λ-point to about 1·3 K. At first the observations gave no indication of quantized circulation, although the values were scattered round h/m_4. Vinen's explanation was that, in the majority of cases, a superfluid circulation existed round only a fraction of the wire. This is permissible provided that a vortex line leaves the wire at some point (Fig. 4.5). The Magnus force changes its value discontinuously at this point, with the result that the measured circulation is an 'average value' for the whole wire. The point of attachment should be fairly mobile, suggesting that states with these non-quantized values of circulation are unstable. In contrast, states in which a whole vortex is attached to the wire should be comparatively stable. To check this hypothesis, the wire was subjected to large-amplitude vibrations prior to the measurement of the circulation, and these results are shown as a histogram in Fig.

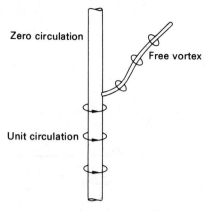

FIG. 4.5 Vortex attached to the wire in Vinen's experiment.

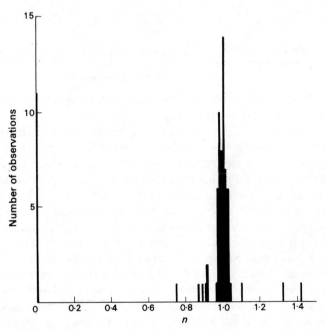

FIG. 4.6 Histogram showing frequency of observation of circulation nh/m_4 plotted against n. (Vinen, 1961.)

4.6. The tendency for values to cluster round one unit of h/m_4 is clearly shown. It was also possible to remove a vortex completely from the wire (values at $\kappa = 0$), and there were still a few cases of partial attachment. Whitmore and Zimmermann (1965) have repeated Vinen's experiment using thicker wires and have detected values of circulation corresponding to 1, 2 and 3 quanta. Furthermore, their

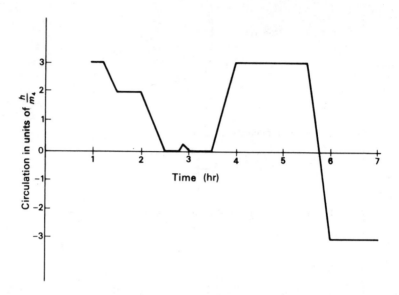

FIG. 4.7   Schematic showing the length of time spent by circulation at various values. System spends some time in non-quantized states, but only quantized states are stable. (Adapted from Whitmore and Zimmerman, 1965 by Putterman, 1972.)

results showed the remarkable stability of quantized circulation, and that non-quantized states are unstable (Fig. 4.7).

The experiments which demonstrated the quantization of circulation, or flux quantization, in superconductors used specimens in the shape of a hollow cylinder. These were made by electroplating or vacuum deposition of the superconducting metal onto a very fine filament. The cylinder was cooled down through its critical temperature in a small uniform magnetic field H along its axis; the field was then switched off; and the flux Φ trapped in the cylinder was measured. In the first experiments (Deaver and Fairbank, 1961, Doll and Näbauer, 1961), Φ was measured by a ballistic method; the more recent experiment of Goodman et al. (1971) made use of a SQUID magnetometer such as we describe in § 5.3. Results of Goodman et al. are shown in Fig. 4.8, and it can be seen that for most values of H, Φ has one of the values $n\phi_0$. There are significant ranges of H, however, for which Φ has some intermediate value. It is plausible that these intermediate values arise for the same reason as do intermediate values in Vinen's experiment, that a flux line is detached from the cylinder at some point along its length. Goodman et al. were able to confirm this conjecture by using a small pickup coil to measure the value of flux trapped as a function of position along the cylinder. As can be seen from Fig. 4.9, the flux contained at any point along the cylinder is indeed an integral number of quanta.

Equations (4.10) and the simpler form (4.13) are of fundamental importance. The flux quantum involves $2e$, the charge on a Cooper pair, and results like those of Fig. 4.8 are the experimental confirmation that the superconducting state is

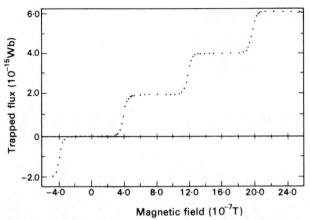

FIG. 4.8 Trapped flux as a function of the magnetic field in which the cylinder was cooled below its transition temperature. These data were taken with a tin cylinder 56 μm inner diameter and 24 mm long, with walls about 500 nm thick. (After Goodman *et al.* 1971.)

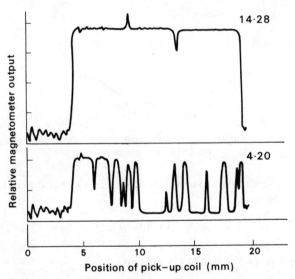

FIG. 4.9 Magnetometer output as a function of position of the pickup coil along the length of the cylinder. The curves are labelled with the values (in units of 10^{-7} Tesla) of applied field in which the cylinder was cooled through its transition temperature. (After Goodman *et al.*, 1971.)

built out of Cooper pairs. We shall see later in this chapter that the flux quantum is important in the mixed state, which consists of quantized vortex lines, and in the next we shall see the part it plays in the understanding of the Josephson effect.

4.2 Quantized vortices in He II

In this section we shall concentrate on vortices in He II; we shall deal with quantized vortices in superconductors in §4.3.

4.2.1   *Rotation of the superfluid*

When the two-fluid model of He II was first suggested, it was generally believed that it would be rather difficult to set the superfluid fraction into rotation because superfluid flow was characterized by the irrotationality condition

$$\operatorname{curl} \mathbf{v}_s = 0. \tag{4.16}$$

To see the implications of eq. (4.16), we consider He II contained in a cylindrical bucket (Fig. 4.10), and find the circulation (eq. 4.3) round a contour such as L_2. By

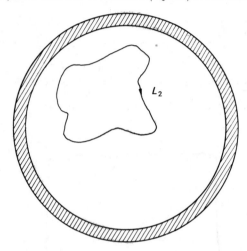

FIG. 4.10   Superfluid in cylindrical container. Contour L_2 in a simply connected region can be reduced to a single point.

Stokes' theorem, the circulation can be written as an integral over the surface A enclosed by the contour:

$$\kappa = \oint_{L_2} \mathbf{v}_s \cdot d\mathbf{l} = \int_A (\operatorname{curl} \mathbf{v}_s) \cdot d\mathbf{A} \tag{4.17}$$

and, combining eqs. (4.16) and (4.17) we find

$$\kappa = \oint_{L_2} \mathbf{v}_s \cdot d\mathbf{l} = 0, \tag{4.18}$$

indicating that the circulation for any contour in the continuous fluid is zero. In addition, it is permissible to reduce all contours to a single point, since the fluid occupies a simply connected region. It follows that eq. (4.18) holds throughout the superfluid only if $\mathbf{v}_s$ is zero at every point. Thus, if eq. (4.16) is true everywhere in the superfluid, rotation is not possible.

This view was strengthened by the results of Andronikashvili's experiment (Fig. 1.7), in which the oscillating pile of disks entrained the normal fluid causing it to rotate, but left the superfluid at rest. However, Osborne (1950) rotated a cylindrical bucket containing He II (Fig. 4.10) and found that the meniscus had the same shape as that of an ordinary liquid undergoing *rigid-body* rotation, which indicated that the superfluid moved at the same angular velocity as the normal fluid. One simple explanation of Osborne's experiment might have been that superfluidity is destroyed when He II is given an angular velocity, but Andronikashvili and Kaverkin (1955) showed that the fountain effect is still present in rotating He II, demonstrating that the distinction between the two fluids is maintained.

The rotation of the superfluid can be satisfactorily explained by assuming that it is threaded by a series of vortex lines (Fig. 4.11). In §4.1 we described how it is

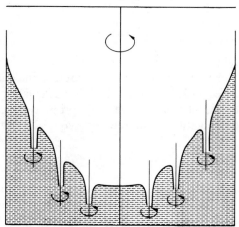

FIG. 4.11   Array of vortex lines in rotating He II.

possible to have a circulation surrounding a region from which the superfluid is excluded. Contours which enclose a solid obstacle or a vortex core (like L_1 in Fig. 4.1) yield a quantized circulation

$$\kappa = \oint_{L_1} \mathbf{v}_s \cdot d\mathbf{l} = \int_A (\text{curl } \mathbf{v}_s) \cdot d\mathbf{A} = nh/m_4. \tag{4.19}$$

Equation (4.19) implies that curl $\mathbf{v}_s$ takes non-zero values at some points of the area A enclosed by the contour. This has no significance for a solid obstacle, but in the case of a vortex we can use it to define what we mean by the core, namely that region in which curl $\mathbf{v}_s \neq 0$.

Thus, although the superfluid in Osborne's experiment appears at first sight to occupy a simply connected region, in which curl $\mathbf{v}_s$ vanishes at all points, it is actually multiply connected owing to the presence of vortex lines, at whose cores curl $\mathbf{v}_s \neq 0$. The way in which rigid-body rotation is simulated will be described shortly when we have looked at the characteristics of vortex lines in more detail.

4.2.2   *An isolated vortex line*

Consider a single straight vortex line in the superfluid (Fig. 4.2). Until §4.2.4 we shall ignore the presence of the normal fluid, so, effectively, we are treating He II at absolute zero. The pattern of flow in the vortex is a series of concentric circular streamlines, lying in planes normal to the axis of the vortex, and surrounding the central cylindrical core. To find how the linear velocity varies with distance from the axis, let us calculate the circulation using a streamline of radius r as the contour. From eq. (4.7) we obtain

$$\kappa = \oint \mathbf{v}_s \cdot d\mathbf{l} = 2\pi r v_s(r) = n \frac{h}{m_4}, \tag{4.20}$$

which leads to

$$v_s(r) = \kappa/2\pi r. \tag{4.21}$$

This dependence of velocity on $1/r$ is the same for a vortex in a classical liquid; the only difference between He II and the classical liquid is the quantization condition. It is obvious from eq. (4.20) that the value of the circulation is independent of the particular streamline chosen, so we may properly speak of the vortex possessing quantized circulation. Also, rewriting eq. (4.20) in the form

$$m_4 r v_s(r) = n\hbar, \tag{4.22}$$

we see that the angular momentum per particle is quantized in units of $\hbar$.

At large values of r the flow pattern associated with a vortex is limited in extent by the boundaries of the liquid or by the presence of other vortices. As $r \to 0$, eq. (4.21) predicts a divergence, which is further evidence that the properties of the core region are different from those of the surrounding liquid. We postpone discussion of the vortex core structure until §4.2.5. For the present, we avoid the divergence by assuming that $\rho_s \to 0$ as $r \to 0$, and that ρ_s falls from its bulk liquid value to zero in a typical distance a_0, which we define as the core radius (Fig. 4.12).

It is worth pointing out that, in introducing the concept of quantized vortices, we have not adhered to the original approach of Feynman (1955). In fact, the earliest suggestion that vorticity should play an important role in superfluid hydrodynamics is credited to Onsager, although he probably thought in terms of vortex sheets rather than lines (see London, 1954). For comparison, we note that Feynman considered the wave function of a ring of atoms in the superfluid, and the effect of allowing each atom to move one place round the ring into the position previously occupied by its neighbour. The final quantum state, following this displacement, is in no way distinguishable from the initial state, because the atoms are Bose particles. Consequently, the wave function is unaltered, and its phase can change only by an integral multiple of 2π. Once again, the phase change is related to the circulation and the quantization of circulation follows directly.

The energy associated with a single vortex line is mainly the kinetic energy of the circulating superfluid. For unit length of line this is

$$\varepsilon_v = \int_{a_0}^{b} \pi \rho_s v_s^2 r \, dr \tag{4.23}$$

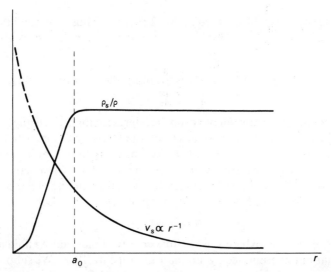

FIG. 4.12   Variation of v_s and ρ_s/ρ with distance r from the axis of a vortex line in He II; a_0 is the core radius (schematic).

and using eq. (4.22) the energy per unit length of a vortex with circulation κ is

$$\varepsilon_v = \frac{\rho_s \kappa^2}{4\pi} \ln\left(\frac{b}{a_0}\right). \tag{4.24}$$

The upper limit in the integral, b, is of the order of the mean separation between vortices, whilst the lower limit is the core radius a_0 (see Problem 4.1). There may be a small extra contribution to the energy arising from distortion of the superfluid wave function at the vortex core, but it is small (Problem 4.2), and is therefore omitted for simplicity.

4.2.3   Vortex lines in rotating He II

We return now to Osborne's rotating-bucket experiment, and explain how the presence of a uniform array of vortex lines enables the superfluid to undergo rigid-body rotation. Suppose there are n_v vortex lines per unit area of the bucket, all having their cores parallel to the axis of rotation (Fig. 4.11), and each one possessing circulation κ. The strength of the array is specified by the vorticity $\bar{\omega}$, defined to be equal to the total circulation within unit area, that is

$$\bar{\omega} = n_v \kappa. \tag{4.25}$$

From eq. (4.17) we see that $\bar{\omega}$ can be identified with the average value of curl v_s. As we have indicated already, when curl v_s takes a finite value it signifies the presence of vortices. The hydrodynamic behaviour of the superfluid as a whole is then described by Euler's equation in its full form (§2.3). In the bucket the total circulation enclosed by a contour of radius R centred on the axis is $\pi n_v R^2 \kappa$. In order for the superfluid to appear to rotate with uniform angular velocity Ω, the total circulation must also be equal to $2\pi R(R\Omega)$. Thus, the condition for the simulation of rigid-body rotation is that the vortex line density is given by

$$n_v = 2\Omega/\kappa. \tag{4.26}$$

Alternatively, we can write this as determining the required vorticity

$$\overline{\omega} = \text{curl } \mathbf{v}_s = 2\mathbf{\Omega}. \tag{4.27}$$

The number and separation of vortex lines present for a given angular velocity clearly depends on their individual circulations. Using a somewhat simplified version of an argument due to Hall (1960) we now establish that it is energetically more favourable to have an array of lines with small circulation than, say, a single line with large κ. We consider He II rotating with the velocity of the container, $\mathbf{\Omega}$. The condition for equilibrium is that the quantity

$$F' = F - \mathbf{L}\cdot\mathbf{\Omega} \tag{4.28}$$

should be a minimum, where F is the free energy of the rotating fluid and $\mathbf{L}$ is its total angular momentum (see e.g. Landau and Lifshitz, 1968). We suppose that the liquid is at a temperature low enough to ignore the difference between ρ_s and ρ, so effectively we are looking at the ground state of rotating He II. If the radius of the container is R_0, the total number of vortex lines with circulation κ is

$$N = 2\pi R_0^2 \Omega/\kappa \tag{4.29}$$

by eq. (4.25). To minimize F' we need consider only those contributions which depend upon N and κ. Since the vortices all have the same circulation each one will have energy ε_v given by eq. (4.24), and the same angular momentum $\mathbf{l}_v$ whose direction is parallel to $\mathbf{\Omega}$. Thus we have to minimize the quantity

$$F'' = N(\varepsilon_v - l_v\Omega). \tag{4.30}$$

A straightforward calculation (Problem 4.3) shows that each vortex contributes angular momentum

$$\mathbf{l}_v = \frac{\rho_s \kappa}{8\pi\Omega}\,\mathbf{\kappa}. \tag{4.31}$$

This is additional to the angular momentum possessed by the fluid by virtue of its rigid-body rotation. Substituting from eqs. (4.24), (4.29) and (4.31) we find

$$F'' = \frac{1}{2}\rho_s R_0^2 \Omega\kappa\left\{\ln\left(\frac{b}{a_0}\right) - \frac{1}{2}\right\}. \tag{4.32}$$

For fixed Ω it follows that κ must take its minimum value, and so the ground state of rotating He II contains a regular array of vortex lines all having the smallest possible circulation h/m_4, which means that the total number of lines present is a maximum. The equilibrium situation is illustrated in Fig. 4.13, which shows the vortex array as seen from a frame rotating at the same speed as the container.

By a simple extension of the argument we have just used, we now estimate the critical value of angular velocity, Ω_{c1}, which must be reached before it becomes possible to create even one vortex. Again equilibrium is determined by the minimum value of the free energy F'. A vortex line will be formed in the rotating

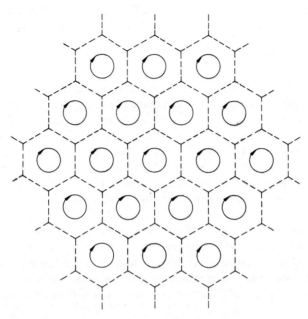

FIG. 4.13 Diagram of the flow due to a close-packed array of vortex lines as seen from a reference frame rotating with angular velocity $\Omega = n_v \kappa/2$, where n_v is the number of lines per unit area. The solid lines are streamlines, while the broken lines are lines of zero velocity in the rotating frame. (Hall, 1960.)

fluid provided that its appearance is accompanied by the reduction of F' below its value with no vortex lines present. This occurs when the quantity $\varepsilon_v(R_0) - l_v(R_0) \cdot \Omega$ becomes negative, where $\varepsilon_v(R_0)$ is the energy and $l_v(R_0)$ the angular momentum of a vortex whose associated flow pattern fills the entire container. The energy is given by eq. (4.24),

$$\varepsilon_v(R_0) = \frac{\rho_s \kappa^2}{4\pi} \ln\left(\frac{R_0}{a_0}\right) \tag{4.33}$$

and we find the angular momentum in the following way:

$$l_v(R_0) = \int_{a_0}^{R_0} \rho_s r v_s (2\pi r)\, dr. \tag{4.34}$$

Using eq. (4.21), and with $R_0 \gg a_0$, we find

$$l_v(R_0) = \tfrac{1}{2}\rho_s \kappa R_0^2. \tag{4.35}$$

The critical velocity for the formation of one vortex with minimum circulation is therefore

$$\Omega_{c1} = \frac{h}{2\pi m_4 R_0^2} \ln\left(\frac{R_0}{a_0}\right). \tag{4.36}$$

If R_0 is given a typical value such as $1 \, \text{cm}$, $\Omega_{c1} \sim 10^{-3} \, \text{s}^{-1}$, showing how easy it is for vortex lines to appear, and conversely that it is difficult to prepare a specimen of He II from which vorticity has been excluded.

However, for very low angular velocities, $\Omega < \Omega_{c1}$, the stable state of the superfluid has $v_s = 0$ everywhere. This state has been observed by Hess and Fairbank (1967) who measured the angular momentum of a cylindrical vessel containing liquid helium as it was cooled through the λ-point. The angular momentum acquired by the superfluid as it was formed, was determined by reheating the cylinder through the λ-point and noting the change of angular momentum as the He I commenced rigid-body rotation. For sufficiently low values of angular velocity, it was found that the superfluid formed in a state of zero angular momentum relative to the laboratory. The threshold value of angular velocity above which the superfluid began to rotate was in good agreement with eq. (4.36).

By an extension of the calculation leading to eq. (4.36), Hess (1967) has found critical angular velocities for the formation of $2, 3, 4$, etc. singly quantized vortices. Thus, a plot of the number of vortices present against angular velocity shows a step structure, and this has been observed by Packard and Sanders (1969), who made use of the trapping of electrons by vortex cores to count the vortex lines (Fig. 4.14).

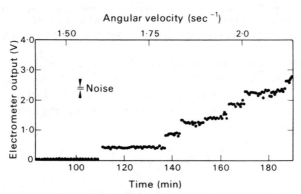

FIG. 4.14 Step structure showing the formation of singly quantized vortices in rotating He II, which is slowly accelerated. The number of vortices present is counted by passing electrons through the liquid, some of which are trapped in the vortex cores; these are subsequently removed by an electric field along the cores. Assuming that vortex traps electrons at a constant rate, the charge collected in fixed time is proportional to the number of vortices present. (Packard and Sanders, 1969.)

As the angular velocity of the container increases further, the density of lines becomes greater (eq. (4.26)), but this cannot continue indefinitely. A limit is set on the proximity of vortex lines by the overlapping of their cores. Using eq. (4.26) again, we see that this would happen if the angular velocity were to reach the higher critical value

$$\Omega_{c2} = \hbar/m_4 a_0^2, \qquad (4.37)$$

which is of the order of $10^{12} s^{-1}$. In all practical situations angular velocities satisfy the condition $\Omega \ll \Omega_{c2}$.

Finally in this section we note the elastic properties of vortex lines. Since the superfluid velocity increases as the core of a vortex is approached, it follows from Bernoulli's equation that there is a corresponding reduction in pressure. In consequence a vortex line is in a state of tension; if it became curved, there would be a restoring force tending to straighten it, this force being equal to the energy per unit length. The process of straightening involves motion of the vortex line through the superfluid so that the line experiences the Magnus force (eq. 4.15) as well (Fig. 4.15). The action of the two forces together causes the propagation of

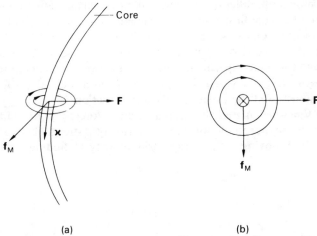

(a)   (b)

FIG. 4.15 (a) Forces acting on a curved vortex line; (b) Section through vortex —the Magnus force acts at right angles to the restoring force, with the result that waves are propagated along vortex.

circularly polarized transverse waves along the vortex line. Hall (1958) has studied these waves by an extension of Andronikashvili's experiment, in which a pile of disks was made to oscillate in rotating He II. The vortex lines stick to the disks, also as a result of the reduced Bernoulli pressure at the core, and the torsional oscillations of the lines are excited by the motion of the disks. Resonances are observed between the two oscillations at frequencies which are in good agreement with theoretically predicted values. The results provide further confirmation of the energy and circulation of vortex lines in He II.

For more details of vortex oscillations, and other features of rotating He II, the reader is referred to the article by Andronikashvili and Mamaladze (1966).

4.2.4 Interaction between vortex lines and the normal fluid

Up to this point our discussion of vortices in superfluid helium has been applicable strictly only at absolute zero and we have totally ignored the normal fluid. However, the presence of vortex lines in the superfluid leads to a force of interaction between the two fluids known as mutual friction, which we mentioned in §2.5. It is the means by which the rotation of the vessel and the normal fluid is communicated to the superfluid to bring about rigid-body rotation (§4.2.3). In addition, mutual friction is important as a factor which limits superfluid velocity. But this is an area of considerable complexity, and we shall give only an outline of the problem.

It appears that the principal source of interaction is the collision of rotons (§1.5) with vortex lines, and most of the experiments have been done in the temperature range where rotons are plentiful, that is, above 1 K. The collision cross-section is clearly a strong function of the direction of the roton drift velocity relative to the vortex lines. If we suppose that there is a uniform array of lines, it is to be expected that rotons moving parallel to the lines will not exchange any momentum with them. This has been confirmed in an experiment by Tsakadze (1962) in which the normal fluid was made to oscillate in a direction parallel to the vortex lines. No difference in the viscous damping of the oscillations could be detected when compared with the same experiment carried out in non-rotating liquid. However, experiments by Hall and Vinen (1956) in which rotons moved at right-angles to the lines demonstrated the existence of mutual friction.

Hall and Vinen calculated the mutual friction force, $\mathbf{F}_{ns}$, on unit volume of normal fluid, due to the interaction between an array of straight lines of vorticity $\bar{\omega}$ and normal fluid moving with average drift velocity $\mathbf{v}_n$, finding

$$\mathbf{F}_{ns} = B \frac{\rho_s \rho_n}{2\rho} \frac{\bar{\omega} \times \{\bar{\omega} \times (\mathbf{v}_n - \mathbf{v}_s)\}}{\bar{\omega}}$$

$$+ B' \frac{\rho_s \rho_n}{2\rho} \bar{\omega} \times (\mathbf{v}_n - \mathbf{v}_s) \qquad (4.38)$$

where $\mathbf{v}_s$ is the average superfluid velocity over a region containing many vortex lines, and B and B' are constants. When the force $\mathbf{F}_{ns}$ is included in the two-fluid eqs. (2.60) and (2.61), the derivation of the equations governing second sound (§2.9) is modified. Qualitatively, since mutual friction appears when there is relative motion between the two fluids and tends to oppose that motion, its effect is to attenuate the second sound. In their experiments, Hall and Vinen used resonant cavities for second sound in which the liquid could be rotated. When the direction of propagation was perpendicular to the vortex lines, extra attenuation was observed of an amount proportional to the angular velocity of the liquid, but second sound propagated parallel to the lines suffered no additional attenuation.

Although these results support the proposition that roton–vortex collisions are the most important factor in mutual friction, the actual nature of the forces experienced by the two fluids is still to be settled. For instance, it is not clear whether the force on a vortex due to roton collisions causes the vortex to move in accordance with the Magnus effect or not. In addition, it has been predicted by Iordanskii (1964) that there will be a force of the form

$$\mathbf{f}_I = \rho_n \mathbf{v}_n \times \mathbf{\kappa} \qquad (4.39)$$

which arises because the normal fluid has to alter its flow pattern to avoid the vortex core. There is a strong similarity of form between the Iordanskii force and the Magnus force, eq. (4.15), but it has been suggested that the former may act on the normal fluid alone. For further discussion of these topics the reader is referred to the papers by Vinen (1966) and Hall (1970).

Landau's two-fluid equations, modified by the addition of the mutual friction force, are still valid only at fairly low velocities. When the fluid velocities become large, for example during the passage of a large heat current, the liquid enters the turbulent regime. Here the vortex lines no longer form a regular array but degenerate into a tangled mass, with the result that the interaction between the fluids is greatly increased.

4.2.5   Ions and vortex rings

Ions have been used frequently as convenient probes to study the properties of He II. The fact that they are charged means that their motion through the liquid can easily be controlled and studied. Usually they are introduced by means of an alpha-active source which ionizes the liquid in its vicinity. The positive ion is probably an agglomeration of neutral atoms plus one α-particle, with diameter between 5 and 10 Å; in contrast, both theory and experiment support the model which pictures the negative ion as a self-trapped electron in a cavity between 12 and 20 Å in diameter (Donnelly 1967). Much information has been obtained from measurements of the ionic mobility

$$\mu_i = \langle v_D \rangle / E \tag{4.40}$$

where $\langle v_D \rangle$ is the average drift velocity and E the applied electric field. The drift velocity is found by observing the ions' time-of-flight between two electrodes placed in the liquid. In the range $0.7\,K < T < 1.8$ K, the mobility of both positive and negative ions is found to vary as $\exp(\Delta/k_B T)$, where Δ is almost exactly the energy of the roton minimum. Remembering that the roton number density N_r is approximately proportional to $\exp(-\Delta/k_B T)$, (§ 1.5), it follows that the ionic mean free path, which is proportional to $\langle v_D \rangle$, varies inversely with N_r. These results indicate that the motion of the ions is governed primarily by collisions with the rotons in this temperature range (for a summary of this work see Careri et al., 1966).

On the other hand, experiments carried out by Rayfield and Reif (1964) near 0.3 K showed that the mean free paths of both positive and negative ions are very much longer than at higher temperatures. Ions having $\langle v_D \rangle$ of the same order as the estimated thermal velocity suffered no energy loss when accelerated by a field over a distance of several centimetres. This result can be understood because at such low temperatures N_r is negligible. However, when the variation of ion energy with velocity was plotted, the energy was found to decrease with increasing velocity (Fig. 4.16). This singular behaviour is characteristic of a vortex ring. Expressions for the kinetic energy and velocity of a vortex ring, with a fluid-filled core undergoing solid-body rotation, are known from classical hydrodynamics

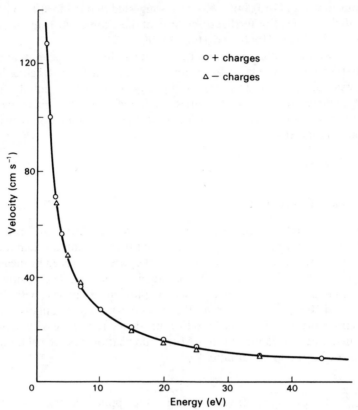

FIG. 4.16   Velocity of ions in He II measured as a function of their energy. The curve is calculated from eqs. (4.41) and (4.42), assuming that an ion is coupled to a singly quantized vortex ring, with core radius $a_0 = 1\cdot3$ Å. (Rayfield and Reif, 1964.)

(Lamb, 1932). Adapting these to a superfluid vortex ring, we have for the energy

$$\varepsilon_r = \tfrac{1}{2}\rho_s\kappa^2 r\{\ln(8r/a_0) - \tfrac{7}{4}\} \qquad (4.41)$$

and for the velocity

$$v_r = (\kappa/4\pi r)\{\ln(8r/a_0) - \tfrac{1}{4}\}, \qquad (4.42)$$

where r is the radius of the ring, κ its circulation and a_0 the core radius, and $r \gg a_0$. Combination of these two equations, neglecting the weak logarithmic terms, indicates that the ring's velocity is approximately proportional to the inverse of the energy (Problem 4.4). In fact, Rayfield and Reif obtained excellent agreement between the experimental data and a curve calculated from the vortex ring equations, by giving the circulation κ the value h/m_4 and assuming that $a_0 = 1\cdot3$ Å (Fig. 4.16). This is a most convincing demonstration that the motion of ions in He II can create vortex rings, which become coupled to the ions. It is also further strong evidence of the quantization of circulation. The radius of the rings in this experiment varied between 5×10^{-6} cm and 10^{-4} cm, which meant that they had an associated flow pattern that was very extensive when compared with the size of the ions.

The mechanism whereby vortex rings are formed in He II is a problem of considerable interest. The view that small rings are created as a result of thermal fluctuations, and that these rings grow to macroscopic size due to collisions with the elementary excitations, was advanced by Iordanskii (1965). This process is similar to the nucleation of a liquid drop in a condensing vapour. The idea has been developed by Donnelly and Roberts (1971), who have made a theoretical study of the motion of ions in He II. As the ion velocity increases there is a growing tendency for excitations to be localized near the ion, and it is suggested that these excitations may include a small vortex ring. Treating the small ring as a Brownian movement particle, and using the standard theory of stochastic processes, these authors calculate the probability P that the vortex ring will grow to its *critical size* when its velocity is equal to that of the ion. After this the ring continues to increase in size until the drag on the ion–ring complex due to the surrounding fluid balances the electric field acting on the ion. The theory suggests that if N ions moving with velocity v_i enter a region where the electric field is constant, the fraction surviving a length L without creating a vortex ring is given by

$$N_i/N = \exp\left(-PN_r L/v_i\right) \tag{4.43}$$

where N_r is the roton density. Strayer and Donnelly (1971) have measured this fraction as a function of velocity, and their results are in excellent agreement with eq. (4.43) (Fig. 4.17).

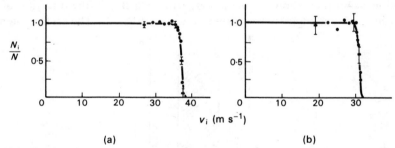

$$\frac{N_i}{N}$$

$$v_i \ (\text{m s}^{-1})$$

(a)   (b)

FIG. 4.17   Fraction of ions surviving for a fixed length in He II in a constant electric field measured as a function of their velocity. (a) Positive ions at 0·7 K and vapour pressure; (b) negative ions at 0·7 K and vapour pressure. Curves are calculated from eq. (4.43). (Strayer and Donnelly 1971.)

The transport of ions through uniformly rotating He II has also been investigated, and in this case there is a marked difference between positive and negative ions. Positive ions behave in a manner similar to that suggested for rotons, appearing to be scattered by the vortex cores. On the other hand, negative ions are trapped in vortex cores, remaining there for a certain characteristic time during which they are able to move along the cores fairly freely. This effect has been demonstrated by Douglass (1964) by passing a transverse negative ion current through rotating He II for a given time. The current was then switched off, and after a delay long enough to allow untrapped ions to be cleared from the liquid, a field was applied parallel to the expected direction of the vortex lines. The charge collected at the top of the vortex array by this means indicated that the negative ions could be trapped in a vortex core with a density as high as 10^7 ions cm^{-3}.

Even from the brief description we have given here, it is clear that the use of ions in He II has proved a powerful technique, especially for the study of vortices. In particular, it provides the best hope for elucidating the structure of the vortex core. A model suggesting that the roton density should be enhanced near the centre of a vortex (Glaberson, 1969) gives a reasonable qualitative description of the motion of negative ions along a vortex core. In addition, measurements of the velocities of vortex rings indicate that the core radius is a slowly increasing function of temperature below 0·6 K (Glaberson and Steingart, 1971). Whether or not a vortex core is hollow, or contains superfluid undergoing solid-body rotation is still an open question.

4.3   Quantized vortices in superconductors

4.3.1   *An isolated vortex line*

For type II superconductors, in which they are found, vortex lines are described by eq. (4.2) together with the Maxwell equation

$$\text{curl } \mathbf{B} = \mu_0 \mathbf{J}_e. \tag{4.44}$$

We recall from § 1.2 that we are using the convention that $\mathbf{J}_e$ is a magnetization current, altering the local value of $\mathbf{B}$.

Suppose that, as in He II, we have a circular flow pattern with a central core. Equation (4.44) then shows that the magnetic field $\mathbf{B}$ is in the direction of the vortex line, at right angles to the current flow —see Fig. 4.18. The vector potential

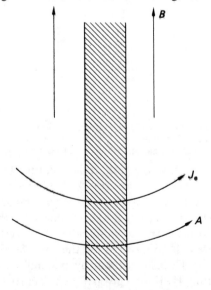

FIG. 4.18   Magnetic field B, supercurrent J_e and vector potential A in a vortex line in a superconductor.

A is at right angles to **B** and we can choose it to lie in the same direction as the supercurrent flow. We introduce the superfluid velocity, as before, by writing

$$\mathbf{J}_e = 2e|\psi|^2 \mathbf{v}_s. \tag{4.45}$$

We can work out the flow pattern by manipulations exactly analogous to those we made for He II. The quantization condition is now the quantization of fluxoid, eq. (4.10).

To find the flow pattern, we eliminate $\mathbf{v}_s$ between eqs. (4.44), (4.45) and (4.10). The resulting equation for B is

$$\lambda^2 \frac{1}{r} \frac{d}{dr}\left(r \frac{dB}{dr}\right) - B = 0, \tag{4.46}$$

where λ is the London penetration depth, given by eq. (3.29). The solution for B is

$$B = \frac{n\phi_0}{2\pi\lambda^2} K_0(r/\lambda), \tag{4.47}$$

in which K_0 is the Bessel function of order zero and imaginary argument and $\phi_0 = h/2e$ is the flux quantum. The associated flow pattern is

$$J_e = -\frac{n\phi_0}{2\pi\lambda^3 \mu_0} K_1(r/\lambda). \tag{4.48}$$

For large distances, $r \gg \lambda$, eq. (4.48) gives

$$v_s \sim \exp(-r/\lambda). \tag{4.49}$$

The flow pattern, and with it B, dies exponentially at large distances. This sharp decay results from the coupling of the current to the magnetic field, and contrasts with the $1/r$ dependence in the He II vortex.

Because of the exponential decay of $\mathbf{v}_s$, we can again simplify the quantization condition, eq. (4.10), for large radius. The integral of $\mathbf{v}_s$ is negligible, and the integral of **A** transforms to a surface integral:

$$\int \mathbf{B} \cdot d\mathbf{\Sigma} = n\frac{h}{2e}. \tag{4.50}$$

Thus the magnetic flux through the vortex line is quantized.

We could now work out the energy of a single flux line, as we did for He II. We shall not do the calculation explicitly; however, we can see immediately that the energy of two separate, single quantum lines ($n = 1$ in eqs. (4.47) and (4.48)) is lower than the energy of a double quantum line ($n = 2$). In fact, the Helmholtz free energy contains a magnetic-field term proportional to B^2 and a kinetic-energy term proportional to J_e^2 (see Problem 4.8). Both of these terms are proportional to n^2, and therefore, as in He II, the energy is minimized for a given **H** by a distribution of single quantum lines. We shall now restrict our attention to a single quantum vortex line.

As in He II, the current, eq. (4.48), diverges for small r, and we must introduce a core radius, which we call $\xi(T)$. We put the flow velocity at radius $\xi(T)$ equal to a critical value v_c; manipulation of eqs. (4.45) and (4.48) and the expression for λ in terms of $|\psi|^2$, eq. (3.27), yields

$$\frac{v_c}{|\psi|} = \frac{e^2 \phi_0 \mu_0^{1/2}}{2^{1/2} \pi m^{3/2}} K_1(\xi/\lambda). \tag{4.51}$$

Obviously the ratio λ/ξ is a fundamental parameter of the superconductor. We use it to define the κ-value:

$$\lambda/\xi = \kappa. \tag{4.52}$$

The picture we now have of a vortex line in a superconductor is summarized in Fig. 4.19(a). The superflow velocity v_s increases as r decreases until it reaches the critical value v_c at the core radius $r = \xi(T)$. We assume that the core of the vortex is normal material, so that $v_s = 0$ for $r < \xi(T)$. Because of the behaviour of v_s, B increases as r decreases, and has a constant value inside the core. Both v_s and B decrease exponentially outside the core, with a characteristic length λ. We note that eq. (4.47) for B outside the core has the property that the total flux of B in the superfluid region is ϕ_0. This form therefore neglects the flux inside the core, so the factor multiplying K_0 is valid only for $\lambda(T) \gg \xi(T)$. Finally, we assumed that $|\psi|$ was constant outside the core region, and the statement that the core is normal implies $\psi = 0$ inside the core. Taking $|\psi|$ to be constant for $r > \xi(T)$ simplifies the calculation, but it is of course open to question. One can study the spatial variation of ψ by using the Ginzburg–Landau (GL) theory, to which we give an introduction in Chapter 6. A GL calculation (Abrikosov, 1957) shows that $|\psi|$ varies continuously from zero at $r = 0$ to be a constant value at $r > \xi(T)$, as sketched in Fig. 4.19(b). From this more general point of view, then, we see that $\xi(T)$ is the characteristic distance for variations in $|\psi|$; whenever $|\psi|$ is varying in space its value changes significantly over distances of order $\xi(T)$. We shall come across more examples of this in Chapter 6. $\xi(T)$ is generally called, confusingly, the coherence length; however, it is important to distinguish $\xi(T)$ from the Pippard coherence length ξ_0, which we introduced in §3.4 as the distance $|\mathbf{r}' - \mathbf{r}|$ out to which changes of the vector potential $\mathbf{A}(\mathbf{r}')$ affect the supercurrent $\mathbf{J}(\mathbf{r})$. One important difference between the two lengths is that ξ_0 is independent of temperature, whereas we shall see that $\xi(T)$ has the same sort of temperature dependence as $\lambda(T)$, and in particular that it diverges at T_c.

We can now see the condition that must be satisfied for quantized vortices to be found in a superconductor. We proceeded by assuming that a vortex did exist, and we found the flow pattern. In particular, the flow has an outer radius $\lambda(T)$ and an inner radius, $\xi(T)$; clearly we need some condition like $\lambda(T) > \xi(T)$ for the vortex to be possible. In terms of the parameter κ, the exact condition is

$$\kappa > 1/\sqrt{2} \tag{4.53}$$

as we shall see in Chapter 6. It is this condition which distinguishes between type I superconductors, in which vortices are not found, and type II superconductors, satisfying the inequality (4.53), in which they are found.

The variation of $\xi(T)$, or κ, with temperature is broadly in line with eq. (4.51). The Landau criterion gives $v_c = \Delta/p$, where p is the momentum of the excitation.

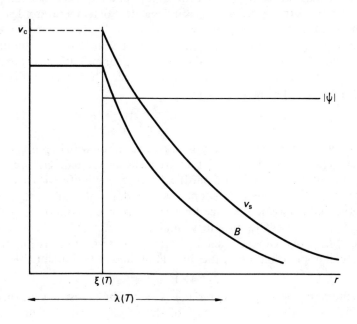

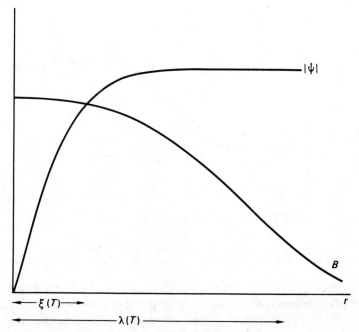

4.19   (a) Wave function ψ, superfluid velocity v_s and magnetic field B as functions of radial distance r in a London vortex. (b) ψ and B as functions of r in a GL vortex.

Recalling from § 1.4 that Δ and $|\psi|$ have the same temperature dependence, we see that eq. (4.51) implies that λ/ξ, and so κ, is independent of temperature. The exact calculation from the microscopic theory does give a weak dependence, κ increasing with decreasing temperature. In the dirty limit κ increases by about 15% between T_c and zero temperature (Maki, 1964, de Gennes, 1964); away from the dirty limit the temperature dependence of κ is stronger. The form of $\xi(T)$ near T_c in the dirty limit is of particular interest:

$$\xi(T) = 0{\cdot}85(\xi_0 l)^{1/2}\left(\frac{T_c}{T_c - T}\right)^{1/2}. \tag{4.54}$$

We may quote from the microscopic theory the variation with impurity content of the parameters we have defined. For a pure metal at zero temperature, $\xi(T)$ is equal to ξ_0, the electromagnetic (Pippard) coherence length of § 3.4. This is not a coincidence: ξ_0 is 'the radius of a Cooper pair', and it is not surprising to see it appear as the characteristic distance for variations in ψ. As impurity content increases, $\xi(T)$ decreases; in the dirty limit $l \ll \xi_0$, and at zero temperature $\xi(T = 0) = (\xi_0 l)^{1/2}$. We can say that ψ 'becomes more flexible' as impurity is added. A useful approximation in the dirty limit, due to Goodman (1962), is

$$\kappa = \kappa_0 + 2{\cdot}4 \times 10^6 \rho \gamma^{1/2}, \tag{4.55}$$

where κ_0 is the value for the pure material, ρ is the residual resistivity in Ω m and γ is the coefficient of the electronic part of the specific heat, in $\mathrm{J\,m^{-3}\,deg^{-2}}$. Equation (4.55) gives good agreement with experimental results —see Fig. 4.20.

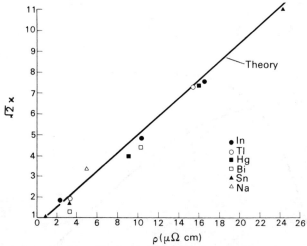

FIG. 4.20   Variation of κ with residual resistivity ρ for a range of Pb specimens containing various impurities. The line marked 'theory' is taken from eq. (4.55), which has no adjustable parameters. (After Livingston, 1963.)

The significance of κ and of $\xi(T)$ becomes clearer if we look at the surface energy of an interface between the superconducting and normal phases. We consider a specimen in an applied field H_{cb}; the Gibbs energy of the superconducting phase with the flux totally excluded is

$$G_s(H_{cb}) = G_s(0) + \tfrac{1}{2}\mu_0 H_{cb}^2, \tag{4.56}$$

which of course is equal to the energy of the normal phase. This means, as we know, that at this particular field value a specimen has the same free energy whether it is in the normal phase or in the superconducting phase with the flux totally excluded. Now suppose that, as in Fig. 4.21, we introduce a plane interface of area A, with the specimen superconducting to the left and normal to the right.

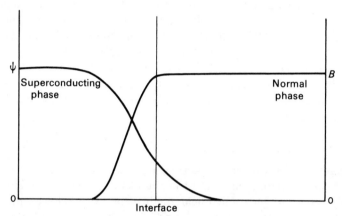

FIG. 4.21   Superconducting–normal interface under applied field H_{cb}.

As sketched in Fig. 4.21, the changeover between the phases is not abrupt: ψ decreases to zero over a distance $\xi(T)$. Hence the superconducting volume is reduced by an amount of order $A\xi(T)$, which leads to a positive contribution of order $(G_n - G_s)A\xi(T)$ to the energy of the interface. On the other hand, the magnetic field penetrates a distance $\lambda(T)$ into the superconducting phase, and this gives a negative surface energy of order $(\mu_0 H_{cb}^2/2)A\lambda(T) = (G_n - G_s)A\lambda(T)$. In all, then, the surface energy per unit area is

$$\sigma = \tfrac{1}{2}\mu_0 H_{cb}^2\{\xi(T) - \lambda(T)\}, \tag{4.57}$$

where both ξ and λ may be multiplied by numerical factors of order unity.

Equation (4.57) shows that for $\kappa = \lambda/\xi$ small, σ is positive, while for κ large σ is negative. Thus type I superconductors have positive surface energy, and type II superconductors have negative surface energy. A more precise calculation, within the framework of the GL theory (see de Gennes, 1966, p. 207), shows that the surface energy changes sign at $\kappa = 1/\sqrt{2}$. The situation is therefore that for $\kappa < 1/\sqrt{2}$ we have positive surface energy and type I behaviour, while for $\kappa > 1/\sqrt{2}$ we have negative surface energy and type II behaviour.

The link between negative surface energy and type II behaviour is easily understood if we again consider a specimen in an applied field H_{cb}. If the specimen is in the Meissner state, the free energy density is $G_s + \tfrac{1}{2}\mu_0 H_{cb}^2$, if it is in the normal state the free energy density is G_n, which is equal to $G_s + \tfrac{1}{2}\mu_0 H_{cb}^2$. However, if we introduce an interface, the energy is lowered from G_n because of the negative surface energy. Thus the total energy will be minimized by the introduction of as many normal-superconducting interfaces as possible, and this is what is achieved by the vortex structure of the mixed state. We can also see from this argument that type I behaviour requires a positive surface energy; as we mentioned in § 3.4 this was one of Pippard's reasons for the existence of a coherence length in superconductors.

We are now in a position to understand the magnetization curves of type II superconductors, which were described in § 1.2. We recall that the magnetization can be either reversible or irreversible. In the former case, as the magnetic field is increased from zero, the flux is completely excluded until a well-defined field H_{c1} is reached; above H_{c1} the flux penetrates the specimen, and the return to the normal state is at the upper critical field H_{c2}. The irreversible curve starts with flux exclusion, has no well-defined lower critical field, and a return to the normal state at H_{c2}. The magnetization in the irreversible case follows a different path when the field is reduced from H_{c2}.

4.3.2   Reversible type II superconductors

The behaviour of a reversible type II superconductor in an increasing applied field is very similar to the behaviour of a container of He II for increasing rotation speeds. For a low enough applied field, the free energy G_s is lower than could be achieved with the partial flux penetration of the mixed state, so the specimen remains in the Meissner state. At H_{c1}, which is analogous to Ω_{c1}, the flux penetrates in the form of vortex lines. As we saw, the lines have an outer radius of about $\lambda(T)$; hence the density of lines just above H_{c1} corresponds to a spacing of $\lambda(T)$, as sketched in Fig. 4.22(a), and the magnetization is sharply decreased from its value in the Meissner state. As H is increased above H_{c1} the flux lines are

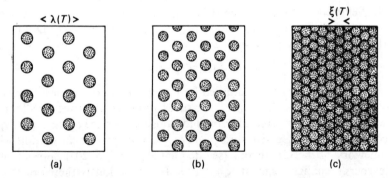

(a)   (b)   (c)

FIG. 4.22   Flux-line distribution (schematic) in a reversible type II superconductor for three different values of applied field H. (a) $H = H_{c1}$, (b) $H_{c1} < H < H_{c2}$, (c) $H = H_{c2}$.

compressed closer together, so that interactions between them are important. The interactions have the consequence that the flux lines lie on a regular lattice. Finally, at H_{c2} the lattice spacing is about $\xi(T)$, and the normal cores of the vortices overlap. As we shall show in Chapter 6, the exact relation between H_{c2} and $\xi(T)$ is

$$\mu_0 H_{c2} = \phi_0/2\pi\xi^2(T). \tag{4.58}$$

Note that at each point on the reversible magnetization curve the flux lines have moved so as to be in the equilibrium state for the applied field, namely a flux lattice of uniform density. This means that a reversible curve will occur only if the flux lines can move freely through the specimen. If the movement of the flux lines is impeded, they will not come into equilibrium, the density of lines will not be uniform, and the magnetization curve will be irreversible.

For He II, we saw that the vortex core is very small, comparable with the interatomic spacing, and the flow pattern extends to the boundaries of the container. In consequence Ω_{c2} is very large and Ω_{c1} is very small. In type II superconductors, on the other hand, the ratio H_{c2}/H_{c1} can have any value from unity upwards; the ratio is governed by the κ-parameter, which from eq. (4.55) is determined by the normal state resistivity.

As we saw in § 1.2, when we have a reversible magnetization curve, we can apply thermodynamics and define a thermodynamic critical field H_{cb} by writing the area under the curve as $\mu_0 H_{cb}^2/2$. The relation between H_{c2} and H_{cb}, which we shall derive in Chapter 6, is

$$H_{c2} = \kappa\sqrt{2}H_{cb} \tag{4.59}$$

which of course has the property that $H_{c2} = H_{cb}$ when $\kappa = 1/\sqrt{2}$.

4.3.3 Irreversible type II superconductors, the Bean model, and flux creep

We have already pointed out that in order for a specimen to show a reversible magnetization curve the flux lines must be able to move freely through the specimen so as to come into equilibrium at every point on the magnetization curve. Conversely, whenever there is any structure in the specimen which impedes the movement of flux lines, the magnetization curve shows some irreversibility. We can again refer to the work of Livingston (1963) for a general view of the basic properties of irreversible materials. Figure 4.23 shows magnetization curves of a lead–indium specimen after successively longer periods of annealing; it can be seen that the degree of irreversibility decreases upon annealing. Note, too, that the

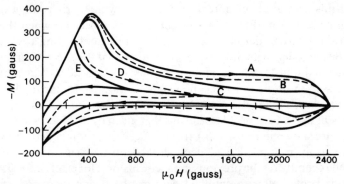

FIG. 4.23 Magnetization curves of a Pb–8·23 wt % In alloy after various periods of room-temperature annealing. A: as cold swaged; B: annealed 30 min; C: 1 day; D: 18 days; E: 46 days. (After Livingston, 1963.)

field at which the normal state is restored, H_{c2}, is independent of the degree of irreversibility. As we have implied, the departures from the reversible curve can be understood in terms of the pinning of the flux lines by metallurgical defects. Thus the flux is prevented from entering abruptly at H_{c1} because it remains pinned near the surface, and the magnetization curve departs only slowly from the initial diamagnetic line. Equally, when the field is lowered from above H_{c2}, flux is trapped in the specimen. The specimen can then be paramagnetic, $\langle B \rangle > \mu_0 H$, rather than diamagnetic, as in some of the curves of Fig. 4.23.

The precise mechanism for pinning is still not fully understood, although it is clear in general terms that structure on the scale of the individual vortex is most effective in pinning. The approach to reversibility upon annealing is evidence that the pinning is in fact due to metallurgical defects. Pinning structure can be introduced by cold working, as in Fig. 4.23, by precipitation of a second phase (Livingston, 1963, for example), or by irradiation (Cullen and Novak, 1964). The flux pinning always increases with increasing inhomogeneity. General reviews are given by Livingston snd Schadler (1964) and by Campbell and Evetts (1972).

The return of the magnetization to zero at H_{c2}, independent of the degree of irreversibility, shows that the strength of flux pinning approaches zero at H_{c2}. This is easy to understand. At H_{c2} the normal cores of the flux lines overlap (eq. 4.58). The interaction between pinning centres and the flux line lattice depends on the difference between the normal cores and the superflow region outside. As H approaches H_{c2} the area of superflow approaches zero, and consequently the pinning strength approaches zero. Thus defects alter H_{c2} only if they alter the mean free path l, and therefore κ.

A model due to Bean (1964) has been very successful in explaining irreversible magnetization curves. Consider the initial portion of the curve for example. In the corresponding reversible curve the flux would enter abruptly at H_{c1}, and just above H_{c1} the flux density across the specimen would be uniform, as illustrated in Fig. 4.22(a). The effect of pinning centres is to hold the flux lines back near the surface; there is a gradient of flux density from the applied value at the surface to zero some way inside, as illustrated in Fig. 4.24. From the Maxwell equation curl $\mathbf{B} = \mu_0 \mathbf{J}$ this field gradient implies a current flowing at right angles to the field. As can be seen from Fig. 4.24, this current arises because when there is a variation in the flux-line density the currents from neighbouring rows do not cancel. The Bean model, in its simplest form, assumes that the effect of pinning is to determine a maximum gradient of flux density, or equivalently a maximum current J_c. As an example, consider the magnetization curve starting from zero applied field and zero magnetization. As the field is increased, penetration starts, and the gradient of the field profile is always that given by J_c. Successive states of the specimen are shown in Fig. 4.25(a). It can be seen that the magnetization of the specimen for a given applied field depends first on the size and shape of the specimen and secondly on the previous magnetic history; for example, Fig. 4.25(b) shows the flux profile after the applied field has been first increased, then decreased. The derivation of the mean induction $\langle B \rangle$ and the magnetization from graphs like Fig. 4.25 is straightforward; an example is given in Problem 4.10.

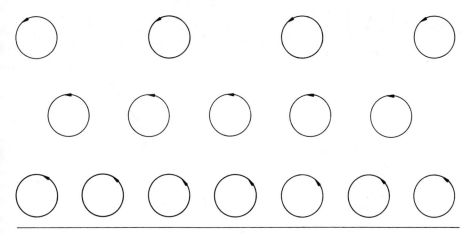

F<small>IG</small>. 4.24   To illustrate the flux-line distribution near the surface when pinning prevents the flux lines
from moving freely into the specimen.

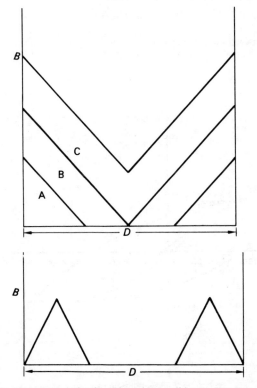

F<small>IG</small>. 4.25   (a) Profiles of magnetic induction B in an irreversible type II superconductor. The specimen
is taken to be a slab of thickness D placed in a parallel applied field, which is increased from zero. The
curves A, B, C represent the profile of B across the specimen for successively larger values of the applied
field, starting from zero. (b) Profile of B after the applied field has been first increased, then decreased to
zero.

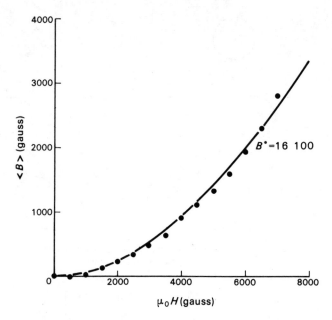

F IG. 4.26   Initial flux penetration in a cylinder of sintered V_3Ga. The curve is theoretical, with an optimal value of the parameter B^*. (After Bean, 1964.)

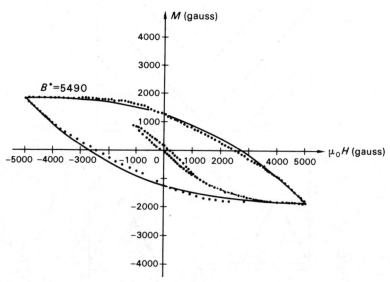

F IG. 4.27   Complete magnetization curve and hysteresis loop for a specimen of lead in porous glass (Vycor). As in Fig. 4.26, the curves are theoretical. (After Bean, 1964.)

The magnetization depends on one adjustable parameter, which in the case of a slab in a parallel field is $B^* = \frac{1}{2}\mu_0 J_c D$. The fit to experimental results is generally very good; examples are shown in Figs. 4.26 and 4.27. It is worth noting that the Bean model has been used in interpreting experiments on low-frequency a.c. losses in irreversible type II superconductors. Through one a.c. cycle, the magnetization traces out a hysteresis loop like the one in Fig. 4.27. The area of the loop, which depends on the peak applied field, gives the loss per cycle. A review of a.c. properties of type II superconductors is given by Melville (1972).

There is no reason to suppose that J_c should actually be independent of the value of B. In fact, we saw that the strength of flux pinning, and with it J_c, tends to zero at applied field H_{c2}, so that J_c must decrease with increasing B. The form $J_c = \alpha/(B + B_0)$, with α and B_0 as adjustable parameters, has frequently been used (Kim et al., 1963).

Finally, we should consider the implications of the fact that the flux distribution at a point on an irreversible magnetization curve is in a metastable state; the flux is held back by the pinning centres present from reaching the equilibrium distribution, which corresponds to a point on the reversible curve. However, the flux lattice has some thermal energy, which means that it is in a state of vibration, and from time to time it may be expected to shake itself free from one pinned configuration, and move to another somewhat nearer the equilibrium one. The first experimental evidence that the flux distribution does change slowly with time was obtained by Kim et al. (1962), who showed in particular that the flux trapped in a hollow cylinder of an irreversible type II superconductor decayed, and that the trapped flux was a linear function of $\ln t$. A later example of this logarithmic time decay is shown in Fig. 4.28. The phenomenon is called flux creep.

A thermal activation process such as we have described commonly leads to a logarithmic time dependence; for example, in some magnetic materials the magnetization increases in this way after a sudden increase in applied field (Street and Woolley, 1949), and as we shall see in § 4.5 the same is true of some superflow patterns in He II.

The theory of flux creep in irreversible type II superconductors was first given by Anderson (1962), and we shall base our account on Anderson's ideas. To begin with, we must go in a little more detail into the mechanism of flux pinning in superconductors, and in particular we can see that the elasticity of the flux line lattice is crucial to pinning (Labusch, 1969). Suppose, for simplicity, that the pinning is due to a random distribution of point centres, as illustrated in Fig. 4.29. It can be seen that the net force on a rigid lattice is zero, since the force on an individual line is as likely to be one way as the other. As soon as one allows for the lattice to have some elasticity, however, one can picture 'bundles' of flux lines piling up behind pinning barriers, and eventually moving on past the barrier because of thermal activation. This admittedly rather vague notion of a flux bundle as the moving quantity was first introduced by Anderson (1962).

Now consider the potential energy $V(x)$ of a flux bundle as a function of position. In the absence of any gradient of flux, the potential is a random series of wells, of average depth U and average width a, say, as in Fig. 4.30(a). In the

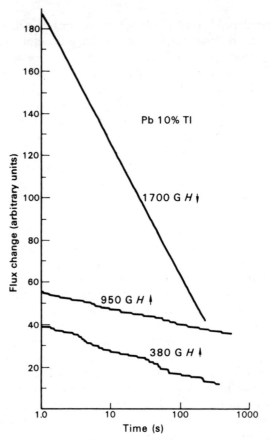

FIG. 4.28   Flux change versus time (logarithmic scale) in a solid cylinder. In each case the value of the applied field is marked, and the arrow indicates whether H was increasing or decreasing. The measurements were taken with a flux transformer and SQUID magnetometer, as described in § 5.3 (After Beasley *et al.*, 1969.)

presence of a flux gradient, as in Fig. 4.25, the potential has an additional uniform slope in one direction, as in Fig. 4.30(b); this slope is caused by the pressure of the uneven flux distribution (Fig. 4.24), or equivalently it is a measure of the Lorentz force exerted on the flux lines by the circulating current $\mathbf{J} = \mathrm{curl}\,\mathbf{B}/\mu_0$. The average slope is

$$(\mathrm{d}V/\mathrm{d}x)_{\mathrm{av}} = BJ lS. \tag{4.60}$$

Here S is the area of the bundle, so that BS is the total flux in the bundle, and we read Fig. 4.30 as the potential energy of a length l of the bundle, where l is the average length of the bundle which interacts with a pinning centre. The flux creep process consists in the hopping of a bundle out of one valley into the next, as indicated in Fig. 4.30(b). Since the valleys have an average width a, the average energy barrier which the bundle has to surmount is $U - BJ lSa$. The flux creep rate R is given by the usual formula for thermal activation:

$$R = \Omega \exp\{-(U - BJlSa)/k_B T\} \tag{4.61}$$

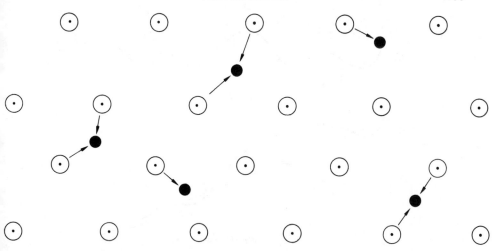

Fɪɢ. 4.29   Interaction of a rigid flux lattice (open circles) with a distribution of point centres (black circles). The centres are taken to be attractive, and the directions of the forces on some individual flux lines are shown.

where Ω is an *attempt frequency*, which we may take as a vibration frequency of the bundle.

We note first that once we admit the possibility of flux creep, the concept of a critical current J_c loses its sharpness: according to eq. (4.61) there is always some movement of flux whatever value J has. However, we can define the critical current J_c in the Bean model as that for which no flux movement is detected within the time of observation, and since R depends exponentially on J, the value of J_c is effectively insensitive to the choice of the time of observation. One important parameter is the value J_{c0} of current at which the average barrier height is zero:

$$J_{c0} = U/BlSa. \tag{4.62}$$

As can be seen from Fig. 4.30(c), for $J > J_{c0}$ there are effectively no barriers for the flux to surmount; in this region one observes flux flow rather than flux creep. We can define the critical current J_c for a given experiment as that for which the creep rate has a critical value R_c, such that rates below that are not observable in that experiment. Clearly

$$J_c = J_{c0}\left(1 - \frac{k_B T}{U} \ln \frac{\Omega}{R_c}\right). \tag{4.63}$$

The experimental results on flux creep show that $U \gg k_B T$ (see Webb, 1971, or Campbell and Evetts, 1972, for example), and consequently the measured value of J_c is J_{c0}. In the particular case of flux lines in a superconductor, therefore, the uncertainty in the definition of J_c is only an uncertainty in principle, and in practice the value is well defined.

We now go on to derive the logarithmic time dependence of the flux in the case of flux creep. The details of the derivation depend upon the geometry chosen, and

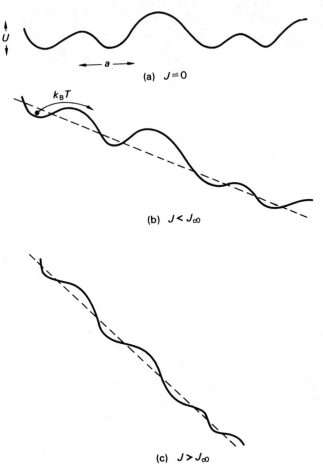

FIG. 4.30   Potential energy of a flux bundle (a) in the absence, (b) and (c) in the presence, of a flux gradient.

for convenience we shall consider the decay of an excess of trapped flux from a hollow cylinder of inner radius r and wall thickness $w \ll r$. The geometry is sketched, and the field profile shown, in Fig. 4.31. With the flux creep rate R, the rate of change of the flux Φ trapped inside the cylinder is

$$d\Phi/dt = -2\pi r B_{in} b R, \tag{4.64}$$

where b is the average distance moved in one flux jump, and we neglect the flux in the cylinder wall. The flux is $\Phi = \pi r^2 B_{in}$ and with the uniform current of Fig. 4.31, we have $\mu_0 J = (B_{in} - B_{out})/w$; substituting these into eq. (4.64) we find the following equation for J:

$$\frac{1}{2}\mu_0 r w \frac{dJ}{dt} = -\Omega B_{in} b \exp\{-(U - B_{in} J l S a)/k_B T\}. \tag{4.65}$$

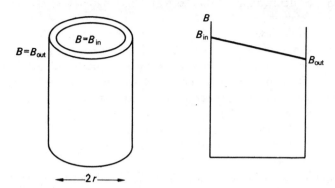

FIG. 4.31   Hollow cylinder with flux trapped, and field profile across wall of cylinder. The field inside is B_{in}, outside is B_{out}, and we assume $B_{in} - B_{out} \ll B_{in}$.

After a sufficiently long time to give $J(t) \ll J(0)$, this has the solution

$$J(t) = J(0) - \frac{k_B T}{B_{in} l S a} \ln t, \qquad (4.66)$$

that is, a logarithmic time dependence such as that shown in Fig. 4.28.

It will be seen that the theory of flux creep which we have outlined contains a number of unknown parameters, such as the barrier height U and the *bundle volume lS*. Measurements of critical current and creep rate, eqs. (4.62) and (4.66), give some information about these parameters. On the other hand, because of the number of parameters, there is still uncertainty even about the fundamental question of whether the creep is thermally activated or driven by some other means. A full discussion of this and other aspects of flux creep will be found in the review article of Webb (1971) and the book by Campbell and Evetts (1972).

4.4   A comparison of He II and superconducting vortices

For convenience we summarize here the resemblances and differences between vortex lines and vortex states in He II and in superconductors.

An individual line obeys a quantization condition. The quantization is of circulation in the He II vortex, and of flux or fluxoid in the superconducting vortex.

A vortex line consists of a central core region surrounded by circulating currents of particles. In superconductors the radius of the core depends upon the mean free path l, and is typically some hundreds of Ångstrom units in size; it may be identified with the temperature-dependent coherence length $\xi(T)$, which represents the distance over which the superfluid wave function falls to zero. As T approaches T_c, $\xi(T)$ diverges. Likewise in He II, the vortex core radius can be identified with the healing length. The evidence from the vortex-ring experiments is that the core radius is approximately equal to the interatomic spacing. By analogy with the superconducting case, a temperature-dependent coherence length can be defined for He II (see §6.8). This length is also estimated to be a few Ångstroms, except near T_λ.

The core of a superconducting vortex comprises normal metal. The core of a He II vortex may be devoid of superfluid, or it may contain superfluid undergoing solid-body rotation; there is some experimental and theoretical evidence that the excitations of the normal fluid tend to cluster near a vortex core.

The flow pattern outside the core varies with distance r as $1/r$ in He II, whilst in a superconductor the coupling to the magnetic field results in a flow pattern dying as $\exp(-r/\lambda)$ for large r. The small core radius and extended flow pattern in the He II vortex mean that it is most closely analogous to a vortex line in an extreme type II superconductor.

Vortex lines are introduced into He II by cooling the liquid in a bucket rotating at angular velocity Ω, and into a superconductor by applying a magnetic field H. The minimum field for the creation of vortex lines, H_{c1}, is that value of H above which the penetration of flux causes a reduction of the free energy. In an exactly similar way one can define the minimum angular velocity for the creation of vortex lines in He II, Ω_{c1}. Equally one can define the upper critical field H_{c2} and the upper critical angular velocity Ω_{c2}, at each of which the vortex cores begin to overlap, resulting in the destruction of bulk superfluidity. Unless $H_{c2} > H_{c1}$ for a given superconductor, the mixed state cannot occur. Since Ω_{c1} is very small and Ω_{c2} is very large, He II for the most part behaves like an extreme type II superconductor.

The shape of the magnetization curve of a type II superconductor depends crucially upon the degree of flux pinning. The rotating bucket of He II is analogous to the reversible mixed state. We shall see later that a superleak for He II is somewhat analogous to an irreversible type II superconductor.

4.5 Dynamics of vortex states

We shall now deal in more general terms with vortex motion and the effect it has on the flow of supercurrents. The central idea is that a moving vortex line always dissipates energy; the dissipation generally appears as a pressure or temperature drop in He II, and as a voltage in a type II superconductor.

4.5.1 *Phase slip*

We can derive an expression for the pressure drop, or voltage, by introducing the notion of phase slip (Anderson, 1966). In Fig. 4.32(a) we show the lines of equal phase for the superfluid circulating an isolated vortex of unit circulation. Suppose this vortex moves right across a trough of He II or a slab of superconductor as in Fig. 4.32(b). Before the vortex moves across we can measure the difference in phase of the superfluid wave function between the ends of the trough along a path like ACB. After the vortex has moved across, we must use a path like AC'B. However, the phase difference is not the same along these two paths; the two results differ by the phase change going once round the vortex line, namely 2π. So if n vortices cross

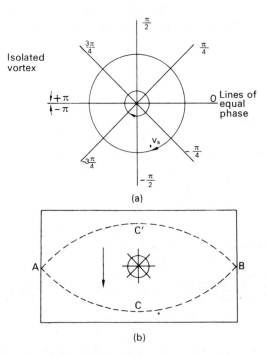

(a)

(b)

FIG. 4.32 (a) Lines of equal phase round an isolated vortex line. (b) Vortex line crossing a channel containing superfluid: phase difference between A and B before vortex crosses (contour ACB) differs by 2π from phase difference after vortex has crossed (contour AC′B).

per second, the phase difference between A and B slips at the following rate

$$\frac{d}{dt}(S_A - S_B) = 2n\pi. \tag{4.67}$$

From Chapter 2 we recall that the superfluid wave function for He II can be expressed as

$$\psi = (\rho_s/m_4)^{1/2} \exp(-i\mu t/\hbar), \tag{4.68}$$

where μ is the chemical potential (eq. 2.16). So long as μ varies slowly with position, the phase in the region of A can therefore be written as

$$S_A = -\mu_A t/\hbar. \tag{4.69}$$

Combining this with eq. (4.67), we find

$$\frac{d}{dt}(S_A - S_B) = -(\mu_A - \mu_B)/\hbar = 2\pi n. \tag{4.70}$$

Equation (4.70) is of fundamental importance. In He II a chemical potential difference can appear either as a temperature difference or as a pressure difference, or both together (eq. 2.37), depending upon the experimental conditions. In either case the difference appears at right angles to the direction of vortex motion. In a superconductor, Gorkov's form of the microscopic theory shows that in equilibrium the condensate wave function has the time dependence $\exp(-2i\mu t/\hbar)$. Here μ is the electrochemical potential, so that 2μ is the energy of an electron pair in the condensate. A difference in electrochemical potential appears as a voltage difference giving

$$2eV_{AB}/\hbar = 2\pi n. \tag{4.71}$$

Again the voltage difference is at right angles to the vortex motion.

Using the time-dependent part of the phase, we can link the vortex motion directly to the loss of energy from the superfluid current. In a situation where there is slow but steady spatial variation in μ, we may take the gradient of the first two terms in eq. (4.70) to obtain

$$\frac{d}{dt}(\hbar \nabla S) = -\nabla\mu. \tag{4.72}$$

In the usual way we make the identification

$$\hbar \nabla S = \mathbf{p}_s, \tag{4.73}$$

where $\mathbf{p}_s$ is the superfluid momentum, so that eq. (4.72) becomes

$$\frac{d\mathbf{p}_s}{dt} = -\nabla\mu = \mathbf{F}, \tag{4.74}$$

which is Newton's Second Law for the superfluid particles. Equations (4.74) and (4.72) are capable of interpretation in two ways. First, the superfluid can be accelerated by an externally applied chemical potential gradient; the relative phases of different points in the fluid will change during acceleration. Secondly, vortex motion across a line joining two points in the superfluid also leads to a chemical potential difference through the phase-slip eq. (4.70), and this causes the energy of the superfluid current between the two points to be dissipated. Of course, the two features of superflow are in no way incompatible, and usually occur together.

In practice, vortices are made to move by a transport current applied at right angles to them. In He II the driving force is the Magnus force, which we introduced earlier (§ 4.1). For superconductors the situation is more complicated. The argument used to derive the Magnus force in He II turns on a transformation to a reference frame in which the normal fluid is at rest. In a superconductor this transformation cannot be carried out, in general, because the metallic lattice provides a fixed frame of reference. The existence or otherwise of the Magnus force in superconductors is therefore a difficult question. Furthermore, because the flow is an electric current, there is a Lorentz force between the current and the magnetic flux lines. This force acts as a driving force on the flux lines. The force on a single line is given by the usual expression

$$\mathbf{F}_L = \mathbf{J} \times \mathbf{\Phi}_0 \tag{4.75}$$

where $\mathbf{J}$ is the transport current, and $\mathbf{\Phi}_0$ is a vector of magnitude $h/2e$ directed along the flux line.

Under the influence of a Magnus force solely, vortices eventually move along with the transport current, because there is then no net force. We may call this motion the Magnus effect. The phase slip is then transverse to the current flow, and in a superconductor, a large transverse (Hall) voltage would appear. On the other hand, under the Lorentz force the vortex motion is at right angles to the current flow, and the voltage is longitudinal. The general case is illustrated in Fig. 4.33. The experimental situation, as we shall see shortly, is that the voltage in a superconductor is predominantly longitudinal, so the Magnus force appears to be unimportant in a mixed-state superconductor. However, it is possible that the Magnus effect, signalled by a large Hall voltage, might appear in an extremely pure superconductor in which the relaxation time to the lattice is very long.

4.5.2 Experimental situation in superconductors

Since dissipation depends on movement of vortices, the transport properties of a superconductor in the mixed state depend on the degree of flux pinning present. Figure 4.34 shows typical results for the longitudinal voltage developed in the slab geometry sketched. Up to a certain critical current I_c no voltage at all is detected. In view of our discussion of flux creep in § 4.3, we can see that in principle however small the current, there must always be some activated movement of flux lines, with a corresponding flux creep voltage. However, the sensitivity of the voltage measurement in experiments like that of Fig. 4.34 is typically about $10^{-5} \, \mathrm{V \, m^{-1}}$, whereas the magnetometer sensitivity in flux creep experiments like that in Fig. 4.28 is equivalent to a voltage sensitivity of about $10^{-14} \, \mathrm{V \, m^{-1}}$. Thus any voltage due to creep does not show up in Fig. 4.34, and we may take it that the critical current I_c corresponds to the current density J_{c0}' of eq. (4.52) at which the average barrier height is zero.

Beyond the curved foot, the origin of which is obscure, on the V–I characteristic, the longitudinal voltage is given by

$$V = R_F(I - I_c). \tag{4.76}$$

This is often called the region of the resistive mixed state. We saw earlier that the measured voltage is at right angles to the direction in which the flux lines move, so that the longitudinal voltage of eq. (4.76) corresponds to a transverse motion of the flux lines. Conversely, measurements of the transverse component of voltage give information about the motion of the flux lines in the direction of the applied current. The results of such measurements (reviewed by Kim and Stephen in Parks, 1969) depend upon the material used, but they always give a very small transverse (Hall) voltage of the same magnitude as that in the normal state. We conclude therefore that in an experiment like that in Fig. 4.34 the flux lines move across the specimen, essentially at right angles to the current.

The flux-flow characteristics of Fig. 4.34 and eq. (4.76) depend on two parameters: the critical current I_c, and the flux-flow resistance R_F. As we have already seen in our discussion of flux creep, I_c is determined by the strength of the flux-pinning centres present; this is illustrated in Fig. 4.34, where the two

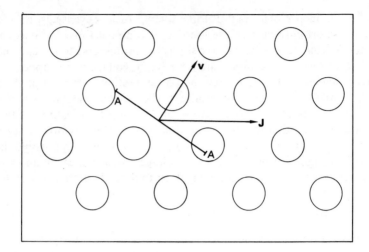

FIG. 4.33 Current flow along **J**, and flux-line movement along **v**. The chemical potential difference (voltage in a superconductor) is along AA, at right angles to **v**. If **v** is along **J** (Magnus effect), the voltage is transverse. If **v** is perpendicular to **J**, the voltage is longitudinal.

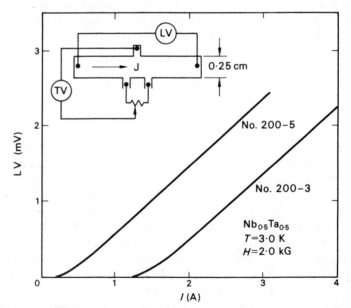

FIG. 4.34 Voltage–current characteristics. The insert shows the geometry of the sheet sample on which voltages in two directions (parallel and perpendicular to J) are measured; the magnetic field is normal to the slab. The lines are $V(I)$ characteristics of two Nb–Ta specimens containing different amounts of defects. (After Strnad *et al.*, 1964.)

specimens have markedly different values of I_c. Once the flux lines are flowing, on the other hand, they reach a terminal velocity which is determined by the balance between the Lorentz force and the viscous drag force on a flux line. This terminal velocity is reflected in the value of R_F. The fact that both specimens in Fig. 4.34 have the same value of R_F is an indication that R_F is relatively insensitive to the nature and strength of the pinning centres, and is determined by properties of the specimen as a whole. In fact, at low temperatures the flux-flow resistivity ρ_F obeys the simple empirical relation

$$\rho_F/\rho_n = \langle B \rangle / \mu_0 H_{c2} \tag{4.77}$$

where ρ_n is the normal state resistivity (Strnad et al., 1964). This equation has a simple interpretation. When the flux lines are pinned and stationary, the transport current is naturally carried as a supercurrent which passes round the normal cores of the flux lines, and no voltage is observed. However, when the flux lines are moving, as in Fig. 4.33, the current cannot pass indefinitely round the cores, and some current must pass through the cores. When eq. (4.77) holds, the current is simply passing straight through the cores, since $\langle B \rangle / \mu_0 H_{c2}$ is a measure of the fraction of the specimen which is occupied by the normal cores.

On their own, results like those of Fig. 4.34 do not show that the voltage in the mixed state is actually due to the movement of flux lines. We may cite two of the more direct forms of evidence that the flux lines do move. First, there are the measurements of voltage noise during flux flow (van Ooijen and van Gurp, 1965, van Gurp, 1968). As we have described it, the voltage is generated by the movement of flux lines from one side of the specimen to the other, as in Fig. 4.33; in view of our discussion of flux pinning we may expect that in general the moving entity is a flux bundle, rather than a single flux line. Because the bundles are discrete, the voltage along the specimen may be expected to fluctuate as a function of time. The simplest view is that the voltage consists of a sequence of superimposed square pulses, each pulse corresponding to the transit of one bundle across the specimen. This is exactly analogous to the voltage generated in an electron tube; here the transport of current is by electrons, and the voltage is a sequence of pulses generated by the individual electrons. Because the voltage is fluctuating in time, it contains components at finite frequencies as well as a d.c. component. In the case of the electron tube the voltage fluctuations are known as shot noise, and the frequency spectrum (mean square voltage $\langle \delta V_f^2 \rangle$ per unit bandwidth) has the form of a single slit Fraunhofer diffraction pattern. Measurements of the noise in the resistive mixed state, as shown in Fig. 4.35, show a form similar to shot noise. If it is assumed that the transit of a single bundle generates a square voltage pulse of duration τ, where τ is the time taken for a bundle to cross the specimen, then the shot noise spectrum in the mixed state is

$$\langle \delta V_f^2 \rangle = 2\phi V_{d.c.} \left(\frac{\sin \pi f \tau}{\pi f \tau} \right)^2 \delta f, \tag{4.78}$$

where ϕ is the bundle size. The transit time τ can be calculated from the d.c. voltage, so the only free parameter in the noise spectrum is the bundle size. It can

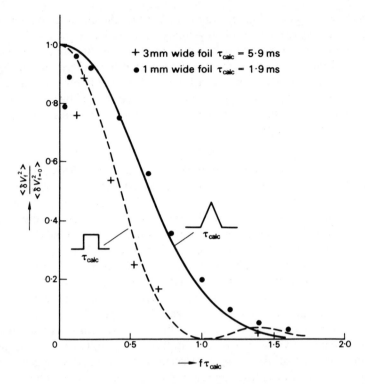

FIG. 4.35 Normalized frequency spectrum of low field noise measured on two V foils 30 μm thick, compared with that calculated for a rectangular (---) and triangular (–) voltage pulse. In each case the transit time τ_{calc} is taken from the d.c. voltage, and the bundle size ϕ is treated as an adjustable parameter. (After van Ooijen and van Gurp, 1965.)

be seen from Fig. 4.35 that the fit of theoretical expressions to the experimental points is good; the bundle size ϕ is often found to be between $10^3\phi_0$ and $10^5\phi_0$. We should add the caution that later experimental results, and their interpretation, are somewhat more complicated than those of Fig. 4.35.

A second piece of evidence that the flux lines move across the specimen in the resistive mixed state comes from measurements of thermomagnetic effects. The core of a flux line consists of normal material, and has some entropy. As a flux line moves across the specimen, therefore, it transports entropy; the side from which the flux lines leave is cooled (removal of entropy) and the other side is heated. In the steady state a temperature difference is developed at right angles to the direction of current flow. The transverse temperature difference ΔT_y is much larger than in the normal state (factor of 10^3), and as is to be expected from the mechanism we have described, it is proportional to the longitudinal voltage ΔV_x. Some experimental results are shown in Fig. 4.36.

A great deal of experimental work has been done on the influence of structural defects on the critical current density J_c. As we have implied, J_c increases as the degree of flux pinning increases. Typical experimental results are shown in Fig.

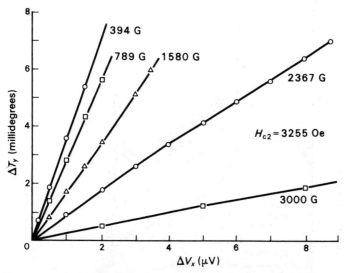

FIG. 4.36   Transverse temperature difference versus longitudinal potential difference for a type II alloy (In–Pb) at various magnetic fields (marked in gauss). (After Solomon and Otter, 1967.)

4.37. As the degree of strain on the specimen is increased, J_c increases. Note that H_{c2} hardly changes, since the residual resistivity ρ, and with it κ, hardly alters during deformation. Fig. 4.37 should be compared with Fig. 4.23, which shows the degree of irreversibility of a magnetization curve increasing as flux pinning increases, again with H_{c2} unchanged. We mentioned earlier various techniques for increasing the degree of flux pinning; further details of the effect of flux pinning on critical currents are given by Livingston and Schadler (1964).

High critical field, high critical current mixed state superconductors have found considerable application in superconducting solenoids, which sustain high magnetic fields without dissipation. Necessary features of a solenoid material are:

(1)   A high critical temperature T_c. This ensures that the thermodynamic critical field H_{cb} is large, and with it $H_{c2} = \kappa\sqrt{2}\,H_{cb}$. In addition, it is hoped to find materials with a high enough T_c for liquid hydrogen to be used as a coolant rather than liquid helium. Existing critical temperatures are slightly too low.

(2)   A high normal state residual resistance ρ. This increases the κ-value, and with it H_{c2}. We recall from § 1.2 that the eventual limit on H_{c2} is set by the Pauli paramagnetism of the normal state.

(3)   A high degree of disorder to give large flux pinning, and therefore a high critical current density J_c. It is obvious that disorder and high ρ go together to some extent.

The situation is slightly paradoxical, as a good solenoid material is a very poor conductor in the normal state. Consequently, if a normal spot appears in a solenoid the local Joule heating is considerable and the normal spot can grow. To guard against such thermal instabilities, a solenoid wire is usually protected by

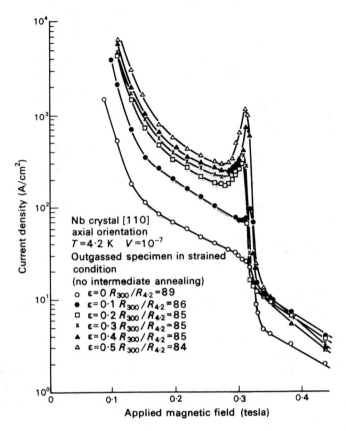

FIG. 4.37   The effect of plastic deformation (bending) on the J_c versus H curve. Transverse applied field. The parameter ε is a measure of the amount of deformation. (After Tedmon et al., 1965.)

copper cladding, the copper providing a low-resistance shunt path round any normal spot that develops.

A review of the extensive literature on superconducting solenoids and their applications is given by Williams (1970).

4.5.3   Vortices and critical velocities in He II

Considerable effort has gone into experiments designed to elucidate the factors which limit the velocity of superflow. As yet there is no complete picture of the processes involved, but it is generally agreed that the production and motion of vortices in the superfluid play a significant part. To explain this point of view here we shall use the results of only a few of the experiments, and in all of these the normal fluid was stationary. Experiments in which the normal fluid is permitted to move as well as the superfluid are naturally more difficult to interpret, since turbulence effects in the normal fluid can easily mask the limiting processes which are intrinsic to the superfluid.

In Chapter 2, we described measurements of gravitational flow through super-leaks made with jeweller's rouge and Vycor, and through saturated and unsa-turated films. In these experiments the variation of the critical volume flow rate σ_{sc} with temperature is determined, where σ_{sc} is conventionally written as

$$\sigma_{sc} = \beta(\rho_s/\rho)v_{sc}d, \tag{4.79}$$

where d is the smallest dimension of the channel, β is a scaling factor appropriate to the particular geometry and v_{sc} is an average *critical velocity*. Unfortunately, it is not straightforward to use eq. (4.79) to find v_{sc}, because in most cases the dimension d is not known accurately. However, using averaged or estimated channel thicknesses, where appropriate, van Alphen *et al.* (1966) found that v_{sc}, defined by eq. (4.79), varies approximately as $d^{-1/4}$, the data coming from measurements made in channels varying from 10^{-7} cm to 10^{-1} cm in width. Thus, v_{sc} appears to be only weakly dependent upon the size of channel.

The first attempt to explain the occurrence of a critical velocity using vorticity was made by Feynman (1955). He calculated the minimum velocity required to create a single vortex ring when regarded as an elementary excitation which, by the Landau criterion, is

$$v_{sc} = (\varepsilon_r/p_r)_{min}, \tag{4.80}$$

where ε_r and p_r are respectively the energy and momentum of the ring. We use eq. (4.41) for the energy, and for the momentum (Problem 4.5)

$$p_r \simeq \pi r^2 \rho_s \kappa, \tag{4.81}$$

with the result that

$$\frac{\varepsilon_r}{p_r} \simeq \frac{\kappa}{2\pi r} \ln\left(\frac{8r}{a_0}\right). \tag{4.82}$$

In a channel of diameter d, the largest possible ring would have radius $\sim d/2$, leading to a critical velocity

$$v_{sc} \simeq \frac{h}{\pi m_4 d} \ln\left(\frac{4d}{a_0}\right). \tag{4.83}$$

This critical velocity depends weakly on the channel size, and predicts that the quantity $v_{sc}d$ should increase slowly with d. Although this behaviour is not in precise agreement with what is found in practice, it is a considerable improvement on the roton critical velocity which is, of course, independent of d. Furthermore, the magnitude of v_{sc} given by eq. (4.83), e.g. 200 cm s^{-1} in a channel with $d \sim 10^{-5}$ cm, is only an order of magnitude greater than measured values. The means whereby a vortex ring is created in a superfluid current are not fully understood, but Glaberson and Donnelly (1966) have described how a straight vortex line, once formed, with both ends pinned to the wall, could distort and expand through the action of the Magnus force, until it produces a ring with the same diameter as the channel (Fig. 4.38).

Evidence that vortex production is a major factor in limiting superfluid flow in films has come from the potentiometer technique of Keller and Hammel (1966),

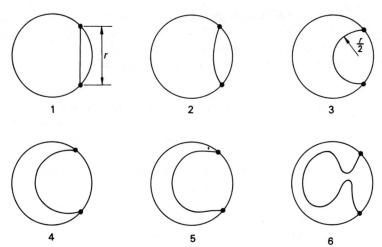

FIG. 4.38   Distortion and expansion of vortex line to form vortex ring (schematic). (Glaberson and Donnelly, 1966.)

which we described in §2.7. We recall that they measured the profile of the chemical potential in the film, finding that its value changes abruptly at the limiting perimeter of the flow path in an outflow experiment, but that it changes more gradually over a considerable length of the film during inflow. In the light of the discussion in §4.5.1, these results can be interpreted as showing that during outflow phase slip occurs only at the limiting perimeter; in other words, the superfluid wave function is coherent throughout the bulk fluid and film on each side of the wall (Fig. 2.8), but there is a discontinuity at the top of the wall. It is also likely that the critical velocity is reached only at the limiting perimeter, and that flow elsewhere in the film is at sub-critical velocities. The conclusion is therefore that, in regions where the critical velocity is reached or exceeded, phase slip occurs, with the implication that these are regions of vortex production. The different behaviour of the chemical potential in inflow experiments signifies that vortex production extends over a considerable fraction of the outer film in that case. We see then that the potentiometer experiments locate the vortex production regions within the film, but a detailed model to show how vortices appear in films is still awaited.

Some progress has been made towards a more general understanding of superfluid critical velocities as the result of a theory proposed by Langer and Fisher (1967). An encouraging feature of the theory is that it ties together several of the ideas such as phase slip and vortex production which, as we have already seen, are very probably linked to critical velocities. However, it breaks with previous approaches to the problem of superfluid stability, by starting from the hypothesis that a steady supercurrent is a non-equilibrium phenomenon. The implications of this proposal have been discussed with reference to a series of experiments on persistent currents in the review article by Langer and Reppy (1970); we shall follow their approach closely in the remainder of this section.

Stable currents of He II can be set up in an annular region as a result of the quantization of circulation (§4.1). These currents have been studied by Reppy and his co-workers using a superfluid gyroscope (Fig. 4.39).

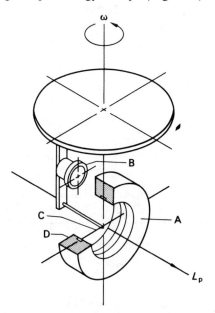

FIG. 4.39   Schematic of superfluid gyroscope. Persistent current with angular momentum L_p is formed in annular container A. Porous material D fills the container. During rotation at ω about the vertical axis, the container deflects against a tungsten fibre C, and the deflection is sensed by detector B. (Langer and Reppy, 1970.)

The annular container is immersed in a pumped He II bath and superfluid enters through a small hole. The container is packed with material that is porous to the superfluid but clamps the normal fluid. A superfluid current is set up by rotating the annulus in the horizontal plane at a fixed angular velocity while cooling the liquid down through the λ-point. At some temperature, the container is brought to rest and then tipped into the vertical plane. The angular momentum vector L_p of the superfluid current is now in the horizontal plane, and it can be made to precess around the vertical axis by the application of a small torque via the horizontal torsion fibre upon which the annulus is mounted. The deflection of the container is detected electronically and this gives a measure of L_p.

For a given pore size, and for each temperature, there is a critical value of L_p which cannot be exceeded however large the initial angular velocity of the container. The temperature variation of L_p(crit) is obtained by setting up the largest possible persistent current, cooling to the lowest available temperature (about 1 K), and then taking measurements of L_p at intervals as the temperature is allowed to rise. The results are shown in Fig. 4.40(a) for 500 Å pores and for Vycor with an average pore size of 40 Å. The critical angular momentum is found to decrease with increasing temperature; this may be due to the reduction in ρ_s/ρ or a

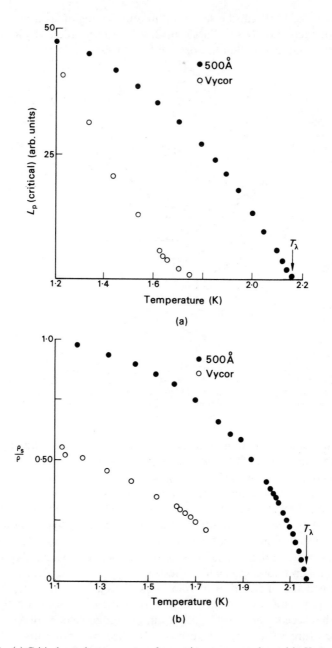

FIG. 4.40   (a) Critical angular momentum for persistent currents formed in Vycor glass and 500 Å filter material as a function of temperature. (b) Relative superfluid density measured in Vycor glass and 500 Å filter material as a function of temperature. Both curves are normalized by using estimates of ρ_s/ρ at 1·1 K for these geometries. (Langer and Reppy, 1970.)

decrease of superfluid critical velocity with increasing temperature, or both together. It is an important feature of the gyroscope method that the variation of ρ_s/ρ with T can be measured for each different pore size used. This is done immediately after the measurements of $L_p(\text{crit})$, which are stopped at a temperature just below the transition temperature. A small reduction in temperature is sufficient to ensure that the superfluid flow velocity is taken below its critical value. Further reduction in temperature now leaves v_s unchanged, but L_p increases again in line with the increase of ρ_s/ρ (Fig. 4.40b). The fact that the angular momentum of a persistent current can be made to increase by simply lowering the temperature is another demonstration of the stability of a macroscopic quantum state.

By subtracting the T-dependence of ρ_s/ρ from that of $L_p(\text{crit})$ one can obtain the behaviour of an averaged superfluid critical velocity (Fig. 4.41). We note that the behaviour of ρ_s/ρ and the critical velocity is very similar to what is seen in flow experiments when a superleak is placed between two reservoirs (Fig. 2.5); in particular, the onset temperature for superflow in very narrow channels is depressed appreciably below T_λ for the bulk liquid, but even in channels only 500 Å wide, the superfluid density takes its bulk value.

In Fig. 4.42, the broken line represents a typical variation of critical velocity with temperature. As we mentioned earlier, if a current flowing at the appropriate critical velocity is set up initially, a reduction of temperature leaves the superfluid velocity unchanged, but now the current flows at a sub-critical value. This change is shown in Fig. 4.42 by the motion of a point, representing the state of the system, along one of the horizontal lines. Each of these lines corresponds to a constant value of v_s and therefore characterizes a stable state of superflow. In a circular ring the superfluid wave function satisfies periodic boundary conditions, with the result that v_s can take only the discrete values $nh/m_4 L$, where n is an integer and L is an average path length round the ring. Consequently the spacing between the horizontal lines (Fig. 4.42) is $\Delta v_s = h/m_4 L$.

Motion of the system point along one of the horizontal lines indicates reversible changes of temperature, during which the superfluid remains in the same metastable state. But, because they are metastable states, there is a finite probability of vertical transitions as well. Langer and Fisher suggest that these transitions will be induced by thermal fluctuations, whose activation energies $E_a(v_s, T)$ are sufficiently great at low T and small v_s to make the fluctuations fairly improbable. Otherwise, it would not be sensible to think in terms of metastable states at all, since their lifetime would be extremely short. Making two further assumptions, that the probability of a fluctuation occurring depends only upon E_a, and that each fluctuation causes the superfluid velocity to decrease by one step Δv_s, it is easy to find an expression for the rate at which the superfluid velocity decays:

$$\frac{dv_s}{dt} = -\frac{hAv_0}{m_4}\exp\{-E_a(v_s, T)/k_B T\}, \tag{4.84}$$

where A is the cross-sectional area of the flow channel, and v_0 is an estimated fluctuation attempt frequency, per unit volume, taken to be approximately equal to the atomic collision frequency times the density.

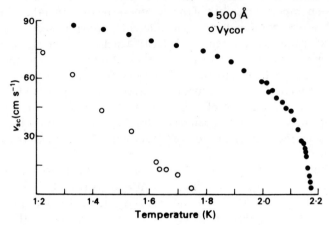

FIG. 4.41   Superfluid critical velocities for flow through Vycor glass and 500 Å filter materials as a function of temperature. Obtained by combining data of Figs. 4.40(a) and (b). (Langer and Reppy, 1970.)

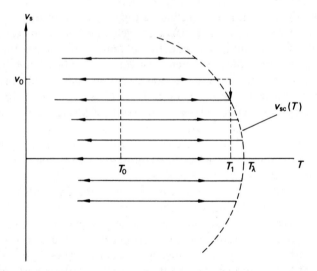

FIG. 4.42   Persistent current-flow velocities (possible metastable states) versus temperature. See text for discussion. (After Langer and Reppy, 1970.)

For consistency with the model described so far, dv_s/dt must take extremely small values at small T and v_s, the region of metastable flow states, but extremely large values at sufficiently high T and v_s. The transition between these two regimes should be very abrupt, occurring in the immediate vicinity of the critical velocity curve. Indeed, this provides a convenient operational definition of v_{sc}, namely that it is the superfluid velocity below which the decay rate dv_s/dt is immeasurably small in the given experimental conditions.

One general prediction of the thermal fluctuation theory has been confirmed in the persistent current experiments. Initially a current is set up at some sub-critical velocity; suppose that this state of the system is represented by the point (v_0, T_0) (Fig. 4.42). The superfluid is then heated until the system point reaches (v_0, T_1), just outside the critical velocity curve. If the temperature is now held constant, the theory predicts that the decay of the superfluid velocity will be logarithmic with time. This behaviour has been observed in a number of different channels and provides strong support for the idea that superfluid states are metastable.

Thus far, the arguments of this theory have been quite general. It is of course equivalent to the theory of flux creep in superconductors, which we discussed in §4.3.3. Note the similarity between eqs. (4.84) and (4.61); and that we derived the logarithmic time-dependence explicitly in the superconducting case (see also Problems 4.11 and 4.12).

To apply the theory more specifically to He II, Langer and Fisher (1967) follow Iordanskii (1965) (§4.2.5) in suggesting that the required thermal fluctuation might be the formation of a small vortex ring. In Fig. 4.43 it is shown how the formation and subsequent expansion of a vortex ring leads to a phase slip of 2π between the ends of the channel. This would correspond to a decrease in the superfluid velocity of Δv_s (Fig. 4.42). Using the standard expressions for the energy and velocity of a ring (§4.2.5), it is possible to estimate the necessary activation energy E_a, and the corresponding temperature-dependent critical velocity.

Experiments have shown that the qualitative features of the model are reproduced in persistent currents, pressure driven superflow (Notarys, 1969) and thermally driven superflow (Liebenberg, 1971), but numerical agreement is poor. In particular, the estimated activation energy is much lower than the experimental value, and correspondingly the measured critical velocities are about an order of magnitude too small. Possible reasons for the discrepancy are that in the non-uniform geometries employed local values of v_s may be anomalously high, and that the activation energy for vortex rings is very probably affected by the proximity of the channel walls. Both of these factors would be extremely difficult to allow for, and almost certainly could not be incorporated in any general theory.

One objection to the theory can be illustrated by reference to superfluid flow through a series of tortuous channels, such as is provided by a superleak made with jeweller's rouge. In this case it is difficult to visualize how a local fluctuation could cause the uniform reduction in superfluid velocity which is apparently observed, unless what is actually measured is the average effect of a series of independent fluctuations. However, these difficulties apart, the fluctuation theory does provide a hopeful approach to the problem of superfluid stability, even if one of its major effects is to render the concept of critical velocity less well defined than had been thought previously.

4.5.4 *The vortex refrigerator*

Vortex motion in He II has been exploited in an ingenious cooling device invented by Staas and Severijns (1969). The principle underlying the working of the refrigerator is familiar from our earlier discussion, namely that a gradient of

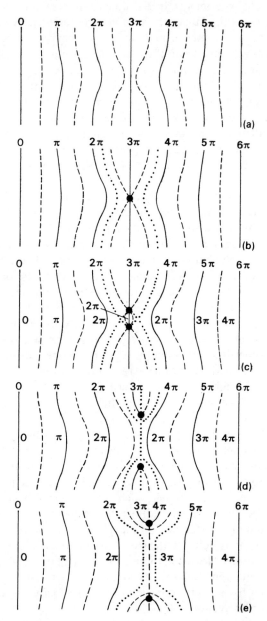

FIG. 4.43   Cross-section of the surfaces of constant phase S, illustrating schematically the successive stages in the nucleation of a vortex ring (c) and its subsequent expansion. Solid lines are surfaces of $S = n\pi$, broken lines $S = (n+\frac{1}{2})\pi$; dotted lines show extra detail. Circles show first vanishing of $|\psi|$ and centres of vortex cores. Cylindrical symmetry about central axis is presumed. Labelling indicates different phase changes along paths drawn through or outside ring. (Langer and Fisher, 1967.)

chemical potential in He II implies the creation and motion of vortex lines. The design of the apparatus is shown schematically in Fig. 4.44; a superleak S, cooling chamber A and capillary C are connected in series, and a fluid pump is used to create a pressure gradient across the combination at constant temperature. The

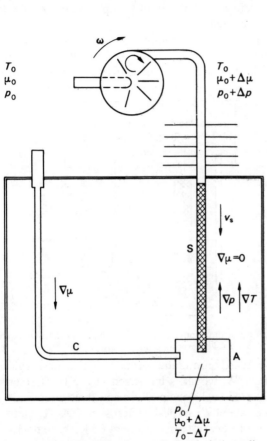

FIG. 4.44   Vortex refrigerator with centrifugal pump. S: superleak, A: cooling chamber; C: capillary.
(Staas and Severijns, 1969.)

high-pressure side is connected to the superleak and the low-pressure side to the capillary. The pump drives superfluid through the superleak at sub-critical velocity, and through the capillary where the critical velocity is exceeded and vortices created as a result.

We may regard the pump as generating a chemical potential difference $\Delta\mu$ across the combination S–A–C. At first, the temperature is uniform, and there is no pressure difference across the superleak. As a result, $\Delta\mu$ appears only in the capillary, associated with vortex production and motion, and is manifested as a pressure difference Δp. The effect of Δp is to drive normal fluid along the capillary as well, and away from the chamber A, which therefore cools down. It can be shown from the equations governing the flow that the superfluid and normal fluid have the same average velocity along C, and that the vortices travel with the two fluids. Consequently $\Delta\mu$ in the capillary remains constant, and so as the temperature gradient in C builds up, Δp diminishes. Eventually a steady state is reached, with a small remaining Δp, and a temperature difference ΔT, whose value depends upon the rate of operation of the fluid pump. Using a centrifugal pump, the chamber A has been cooled to 0·6 K.

A modified version of the refrigerator employs a fountain pump (Fig. 4.45), comprising a second superleak S_2 and a chamber B in which the fluid can be heated electrically. Superfluid passing through S_2 is converted to normal fluid in B

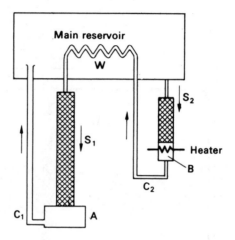

FIG. 4.45 Vortex refrigerator without moving parts. S_1, S_2: superleaks; A,B: liquid reservoirs; C_1, C_2: capillaries. (Staas and Severijns 1969.)

when the heater is switched on, and the normal fluid then passes along capillary C_2 to the heat exchanger W, which is immersed in the He II reservoir. In W the normal fluid is converted back to superfluid and then enters S_1. By this means a chemical potential gradient is established across $S_1 - A - C_1$, and A is cooled as before. Starting with a reservoir temperature of 1·5 K, the temperature of A can be reduced to 0·7 K, significantly lower than can be reached by reducing the helium vapour pressure. However, the greatest advantage of the vortex refrigerator in its second form is the absence of moving parts.

Problems

4.1   Check the derivation of the energy of a vortex line (eq. 4.24).

4.2   Estimate roughly the contribution to the vortex energy from the distortion of the wave function near the vortex core, using the term in $\nabla^2(\sqrt{\rho_s})$ from eq. (2.12). Compare this with the energy of a single vortex (eq. 4.24). Calculate the average vortex separation b using eq. (4.26) with $\Omega = 1\,s^{-1}$; assume that temperature is low enough for $\rho_s = \rho$.

4.3 Show that the angular momentum of a vortex line with circulation κ in He II undergoing solid-body rotation at angular velocity Ω is given by eq. (4.31).

4.4 Show from eqs. (4.41) and (4.42) that the energy of a vortex ring is approximately proportional to the inverse of its velocity.

4.5 Equation (4.41) gives the energy of a vortex ring in a stationary superfluid. If now the superfluid moves with velocity v_s and the vortex-ring velocity v_r (eq. 4.42) is in the upstream direction, the energy of the ring in the laboratory frame becomes $\varepsilon' = \varepsilon_r - p_r v_s$, where p_r is the momentum of the ring, and we have used the Galilean transformation. (a) Assuming that p_r is defined by $v_r = \partial \varepsilon_r / \partial p_r$, and that $p_r = 0$ when the radius of a ring is zero, show that

$$p_r \simeq \pi \rho_s \kappa r^2.$$

(b) Hence show that there is a critical radius r_c at which $\varepsilon'(r)$ is a maximum, and show that

$$\varepsilon'(r_c) \simeq \frac{\rho_s \kappa^3}{16 \pi v_s} \left(\ln \frac{8 r_c}{a_0} - \frac{1}{4} \right) \left(\ln \frac{8 r_c}{a_0} - \frac{11}{4} \right).$$

Once a ring reaches this size, it is energetically favourable for it to grow larger until it is annihilated at the walls of the containing vessel (see discussions in §§ 4.2.5 and 4.5.3).

4.6 It has been pointed out by Roberts and Donnelly (1970) that if eqs. (4.41) and (4.42) are used for a vortex ring in He II, i.e. equations for the case of a fluid-filled core, then the relation $\partial \varepsilon_r / \partial p_r = v_r$, where p_r is defined in Problem 4.5, holds only under the condition of constant core volume. Check that this is so. This result implies that the core radius would, for example, be a function of the momentum of the ring. To avoid this, Roberts and Donnelly consider a ring with a hollow core, and find that its total energy E is made up of two parts, the classical expression for the kinetic energy

$$\frac{1}{2} \rho \kappa^2 r \left(\ln \frac{8r}{a_0} - 2 \right)$$

together with a potential energy $\frac{1}{4} \rho \kappa^2 r$ arising from the fact that the vortex must do work against external pressure when it expands. Show that the expression for the velocity is then

$$v_r = \left(\frac{\partial E}{\partial p_r} \right)_{a_0} = \frac{\kappa}{4 \pi r} \left(\ln \frac{8r}{a_0} - \frac{1}{2} \right),$$

which is the classical expression for the velocity of a hollow-core vortex ring.

4.7 Derive eq. (4.46) using eq. (4.44) and the differential form of eq. (4.10), namely

$$\left(\frac{m}{e |\psi|^2} \right) \text{curl } \mathbf{J}_e + 2e\mathbf{B} = 0,$$

[For $\mathbf{J}_e$ and $\mathbf{B}$ in the directions shown in Fig. 4.18 the components of curl are

$$\text{curl } \mathbf{B} = -\partial B / \partial r \qquad \text{(azimuthal component)}$$

$$\text{curl } \mathbf{J}_e = r^{-1} \partial (r J_e) / \partial r. \qquad \text{(z component).]}$$

The defining equation for the Bessel function of imaginary argument $K_n(x)$ is

$$x^2 \frac{d^2 K_n}{dx^2} + x \frac{dK_n}{dx} - (x^2 + n^2) K_n = 0.$$

The derivatives satisfy

$$dK_0(x)/dx = -K_1(x)$$
$$d\{xK_1(x)\}/dx = -xK_0(x).$$

The asymptotic behaviour of K_1 as $x \to 0$ is

$$xK_1(x) \to 1.$$

Use these properties to derive eqs. (4.47) and (4.48), and show that eq. (4.47) includes the flux quantization condition

$$\int_0^\infty 2\pi r B(r)\, dr = n\phi_0.$$

(This is how the amplitude of B in eq. (4.47) is determined.)

4.8   The Helmholtz free energy in a superconductor is

$$F = E_{kin} + E_{mag}$$

where E_{kin} is the kinetic energy of the supercurrents, and E_{mag} is the usual magnetic field energy. E_{kin} is

$$E_{kin} = \int \tfrac{1}{2} m n_s v_s^2 d^3 \mathbf{r}.$$

Show that F can be written

$$F = (1/2\mu_0) \int \{ \mathbf{B}^2 + \lambda^2 (\text{curl } \mathbf{B})^2 \}\, d^3 r.$$

4.9   Check the derivation of eq. (4.51).

4.10   *The Bean model.* Consider the initial magnetization curve of a slab of thickness D in a parallel applied field, as shown in Fig. 4.25. Show that the induction B satisfies

$$\frac{\partial B}{\partial x} = \pm \mu_0 J_c$$

where the x-axis is normal to the slab, and the choice of sign depends on the direction of the screening current. Hence show that B is zero over some region in the middle of the specimen as long as $B_0 < B^*$, where B_0 is the value of the induction at the surface, and

$$B^* = \mu_0 J_c D/2.$$

Derive the following expressions for the mean induction $\langle B \rangle$ (i.e. the average across the specimen) and magnetization $M = \langle B \rangle - B_0$:

$$\left. \begin{array}{l} \langle B \rangle = B_0^2/2B^* \\[4pt] -M = B_0 - B_0^2/2B^* \end{array} \right\} \quad B_0 < B^*$$

$$\left.\begin{array}{l} \langle B \rangle = B_0 - \tfrac{1}{2}B^* \\ -M = \tfrac{1}{2}B^* \end{array}\right\} \qquad B_0 > B^*.$$

Sketch $\langle B \rangle$ and M as functions of B_0.

4.11 Write down the complete solution of eq. (4.65) for the current $J(t)$ in a flux creep experiment, and confirm that eq. (4.66) is the solution after a sufficiently long time.

4.12 Show that for sufficiently long times, the decay of the superfluid velocity defined by eq. (4.84) shows a logarithmic time dependence.

References

ABRIKOSOV, A. A., 1957, *Zh. Eksper. Teor. Fiz.* **32**, 1442. (*Soviet Physics JETP* **5**, 1174.)

ANDERSON, P. W., 1962, *Phys. Rev. Lett.* **9**, 309.

ANDERSON, P. W., 1966, *Quantum Fluids*, ed. D. F. Brewer (North-Holland, Amsterdam), p. 146.

ANDRONIKASHVILI, E. L. and KAVERKIN, I. P., 1955, *Zh. Eksper. Teor. Fiz.* **28**, 126. (*Soviet Physics JETP* **1**, 174.)

ANDRONIKASHVILI, E. L. and MAMALADZE, YU. G., 1966, *Rev. Mod. Phys.* **38**, 567.

BEAN, C. P., 1964, *Rev. Mod. Phys.* **36**, 31.

BEASLEY, M. R., LABUSCH, R. and WEBB, W. W., 1969, *Phys. Rev.* **181**, 682.

CAMPBELL, A. M. and EVETTS, J. E., 1972, *Critical Currents in Superconductors* (Taylor and Francis, London).

CARERI, G., DUPRE, F. and MAZZOLDI, P., 1966, *Quantum Fluids*, ed. D. F. Brewer (North-Holland, Amsterdam), p. 305.

CULLEN, G. W. and NOVAK, R. L., 1964, *App. Phys. Lett.* **4**, 147.

DEAVER, B. S. and FAIRBANK, W. M., 1961, *Phys. Rev. Lett.* **7**, 43.

DE GENNES, P. G., 1964, *Phys. Kond. Materie.* **3**, 79.

DOLL, R. and NÄBAUER, M., 1961, *Phys. Rev. Lett.* **7**, 51.

DONNELLY, R. J., 1967, *Experimental Superfluidity* (University of Chicago Press, Chicago and London), Chapter 6.

DONNELLY, R. J. and ROBERTS, P. H., 1971, *Phil. Trans. Roy. Soc.*, **A271**, 41.

DOUGLASS, R. L., 1964, *Phys. Rev. Lett.* **13**, 791.

FEYNMAN, R. P., 1955, *Progress in Low Temperature Physics*, ed. C. J. Gorter, Vol. 1, Ch. II (North-Holland, Amsterdam).

GLABERSON, W. I., 1969, *J. Low Temp. Phys.* **1**, 289.

GLABERSON, W. I. and DONNELLY, R. J., 1966, *Phys. Rev.* **141**, 208.

GLABERSON, W. I. and STEINGART, M., 1971, *Phys. Rev. Lett.* **26**, 1423.

GOODMAN, B. B., 1962, *IBM J. Res. Dev.* **6**, 63.

GOODMAN, W. L., WILLIS, W. D., VINCENT, D. A. and DEAVER, B. S., 1971, *Phys. Rev.* **B4**, 1530.

HALL, H. E., 1958, *Proc. Roy. Soc.* **A245**, 546.

HALL, H. E., 1960, *Adv. Phys.* **9**, 89.

HALL, H. E., 1970, *J. Phys. C.* **3**, 1166.

HALL, H. E. and VINEN, W. F., 1956, *Proc. Roy. Soc.* **A238**, 204.

HESS, G. B., 1967, *Phys. Rev.* **161**, 189.

HESS, G. B. and FAIRBANK, W. M., 1967, *Phys. Rev. Lett.* **19**, 216.

IORDANSKII, S. V., 1964, *Ann. Phys. N.Y.* **29**, 335.

IORDANSKII, S. V., 1965, *Zh. Eksper. Teor. Fiz.* **48**, 708. (*Soviet Physics JETP* **21**, 467).

KELLER, W. E. and HAMMEL, E. F., 1966, *Phys. Rev. Lett.* **19**, 998.

KIM, Y. B., HEMPSTEAD, C. F. and STRNAD, A. R., 1962, *Phys. Rev. Lett.* **9**, 306.

KIM, Y. B., HEMPSTEAD, C. F. and STRNAD, A. R., 1963, *Phys. Rev.* **129**, 528.

LABUSCH, R., 1969, *Phys. Status Solidi* **32**, 439.
LAMB, H., 1932, *Hydrodynamics*, 6th edn. (Dover reprint: New York 1965), p. 241.
LANDAU, L. D. and LIFSHITZ, E. M., 1968, *Statistical Physics*, 2nd. edn. (Pergamon Press, London), p. 71.
LANGER, J. S. and FISHER, M. E., 1967, *Phys. Rev. Lett.* **19**, 560.
LANGER, J. S. and REPPY, J. D., 1970, *Progress in Low Temperature Physics*, ed. C. J. Gorter, Vol. VI, Ch. I (North-Holland, Amsterdam).
LIEBENBERG, D. H., 1971, *Phys. Rev. Lett.* **26**, 744.
LIVINGSTON, J. D., 1963, *Phys. Rev.* **129**, 1943.
LIVINGSTON, J. D. and SCHADLER, H. W., 1964, *Progress in Materials Science* **12**, 183.
LONDON, F., 1954, *Superfluids*, Vol. II (Dover reprint 1964, New York), p. 151.
MAKI, K., 1964, *Physics* **1**, 21.
MELVILLE, P. H., 1972, *Advances in Physics* **21**, 647.
NOTARYS, H. A., 1969, *Phys. Rev. Lett.* **22**, 1240.
OSBORNE, D. V., 1950, *Proc. Phys. Soc.* **A63**, 909.
PACKARD, R. E. and SANDERS, T. M., JR., 1969, *Phys. Rev. Lett.* **22**, 823.
PARKS, R. D., 1969, *Superconductivity*, 2 vols. (Dekker, New York).
PUTTERMAN, S. J., 1972, *Physics Reports* **4**, 67.
RAYFIELD, G. W. and REIF, F., 1964, *Phys. Rev.* **136**, 1194.
ROBERTS, P. H. and DONNELLY, R. J., 1970, *Phys. Lett.* **31A**, 137.
RUTHERFORD, D. E., 1959, *Fluid Dynamics* (Oliver and Boyd, Edinburgh), §24.
SOLOMON, P. R. and OTTER, F. A., JR., 1967, *Phys. Rev.* **164**, 608.
STAAS, F. A. and SEVERIJNS, A. P., 1969, *Cryogenics* **9**, 422.
STRAYER, D. M. and DONNELLY, R. J., 1971, *Phys. Rev. Lett.* **26**, 1420.
STREET, R. and WOOLLEY, J. C., 1949, *Proc. Phys. Soc.* **A62**, 562.
STRNAD, A. R., HEMPSTEAD, C. F. and KIM, Y. B., 1964, *Phys. Rev. Lett.* **13**, 794.
TEDMON, C. S., JR., ROSE, R. M. and WULFF, J., 1965, *J. Appl. Phys.* **36**, 829.
TSAKADZE, D. S., 1962, *Zh. Eksper. Teor. Fiz.* **42**, 985. (*Soviet Physics JETP* **15**, 681).
VAN ALPHEN, W. M., VAN HAASTEREN, G. J., DE BRUYN OUBOTER, R. and TACONIS, K. W., 1966, *Phys. Lett.* **20**, 474.
VAN GURP, G. J., 1968, *Phys. Rev.* **166**, 436.
VAN OOIJEN, D. J. and VAN GURP, G. J., 1965, *Phys. Lett.* **17**, 230.
VINEN, W. F., 1961, *Proc. Roy. Soc.* **A260**, 218.
VINEN, W. F., 1966, *Quantum Fluids*, ed. D. F. Brewer (North-Holland, Amsterdam), p. 74.
WEBB, W. W., 1971, *J. Appl. Phys.* **42**, 107.
WHITMORE, S. C. and ZIMMERMANN, W., 1965, *Phys. Rev. Lett.* **15**, 391.
WILLIAMS, J. E. C., 1970, *Superconductivity and its Applications* (Pion, London).

Chapter 5

Josephson Effects

5.1 Introduction

We have already discussed, in Chapter 3, single-particle tunnelling between superconductors separated by an oxide barrier. The insulating barriers used in that kind of experiment are sufficiently thick for the superconducting wave functions on either side not to overlap in the barrier. However, with very thin oxide barriers significant overlap does occur, and a number of new effects, collectively called Josephson effects, are then seen. The simplest consequence of overlap is that a small d.c. supercurrent can flow: the barrier is acting as a *weak superconductor*. Furthermore, if a voltage is established across the junction, there is a corresponding rate of phase slip between the two sides, and this phase slip can synchronize with an applied a.c. electric field, for example. Thus as well as the d.c. effect there is a number of interesting a.c. effects.

Oxide barriers are not the only kind of *weak link* between superconductors: the criteria for Josephson effects to occur are first that the coupling between the superconductors should be strong enough for the phases of the wave functions on either side to be related, and secondly that the coupling should be weak enough for the system to be perturbed by applied radiation fields. Common types of weak link systems are:

(1) An oxide barrier between evaporated thin films. The Al_2O_3 frequently used for single particle tunnelling is not easily made thin enough, and the commonest barrier is the oxide on a Sn or Pb film.

(2) A point contact between two bulk superconductors. Typically, a Nb wire is ground to a point, the surface is allowed to oxidize, and then the point is pressed against a piece of bulk Nb. Point contacts are frequently set up in such a way that the pressure can be adjusted from outside the cryostat.

(3)   A thin film *microbridge* formed by etching down a superconducting film, as sketched in Fig. 5.1.

FIG. 5.1   Thin film microbridge. The dimensions of the constriction are less than 1 μm.

These systems differ in the details of their behaviour, but all show the common Josephson effects. The differences in behaviour are partially correlated with the strength of coupling, which increases in the order oxide—point contact—microbridge. In addition, the parameters of the electrical circuit of which the weak link forms a part have considerable effect.

In principle, Josephson effects should also be observable when two containers of He II are weakly coupled together. In practice, the only type of weak coupling that has been used is a pinhole between reservoirs. Several experimenters, using this arrangement, have obtained results which they believed to be manifestations of an a.c. Josephson effect, but later work has rather undermined this interpretation.

We have developed our treatment of superconductors and He II very much from a 'quantum fluid' point of view, stressing the importance of the macroscopic wave function. The fact that two wave functions can overlap in a barrier region and that a weak supercurrent can then pass may seem rather obvious from that point of view. Historically, however, the prediction of the Josephson effect (Josephson, 1962) and subsequent clarification (Josephson, 1965, Anderson 1964, 1966, 1967) did much to establish our present understanding of superfluid systems. For example, the notion of *phase slip*, on which we based our discussion of the dissipation produced by vortex motion (Chapter 4), was first introduced by Anderson in connection with the a.c. Josephson effect. In addition, the discussion of the Josephson effect led to a clearer understanding of the way in which the transition to the superconducting or superfluid state involves an ordering, analogous say to the alignment of spins in a ferromagnet at the Curie point (see § 6.9).

We derive the governing equations for the Josephson effect in the next section, and in subsequent sections we deal successively with the d.c. and a.c. effects in superconductors, and with the He II analogue. A more detailed account of the Josephson effect is given by Solymar (1972).

5.2   Basic equations

We outline the very lucid derivation given by Feynman (1965). Putting ψ_1 and ψ_2 for the wave functions on either side of the barrier, we write the equations of motion as

$$ih \frac{\partial \psi_1}{\partial t} = \mu_1 \psi_1 + K \psi_2, \tag{5.1}$$

$$ih\frac{\partial\psi_2}{\partial t}=\mu_2\psi_2+K\psi_1. \tag{5.2}$$

Here K represents the coupling across the barrier. In the absence of the coupling, as we know, the wave functions depend on time as $\exp(-i\mu t/\hbar)$, so we have written the diagonal terms as $\mu_1\psi_1, \mu_2\psi_2$ respectively. We solve the equations by putting, analogously to eq. (2.1).

$$\psi_1=n_1^{1/2}\exp(iS_1), \tag{5.3}$$

$$\psi_2=n_2^{1/2}\exp(iS_2), \tag{5.4}$$

where n_1, n_2 are the superfluid number densities, and S_1, S_2 the phases. As we have already seen in discussing eqs. (2.1) and (2.3), to specify both n and S as we are doing requires that the number N of particles in each superfluid is large, since N and S must satisfy the uncertainty product

$$\delta N\,\delta S\sim1, \tag{5.5}$$

so that S can have an uncertainty $\delta S\sim N^{-1/2}$, and N the relative uncertainty $\delta N/N\sim N^{-1/2}$. By analogy, we see that eqs. (5.3) and (5.4) presuppose that the total number of particles in the superfluid phase on either side of the barrier is large.

Substituting eqs. (5.3) and (5.4) into (5.1) and (5.2), we get

$$\hbar\frac{\partial n_1}{\partial t}=2Kn_1^{1/2}n_2^{1/2}\sin(S_2-S_1), \tag{5.6}$$

$$\hbar\frac{\partial n_2}{\partial t}=-2Kn_1^{1/2}n_2^{1/2}\sin(S_2-S_1), \tag{5.7}$$

$$-\hbar\frac{\partial}{\partial t}(S_2-S_1)=\mu_2-\mu_1. \tag{5.8}$$

These are the governing equations for the Josephson effect. Equations (5.6) and (5.7) describe a current flowing across the barrier, while eq. (5.8) is the familiar equation for phase slip; for a superconductor we write $\mu_2-\mu_1=2eV$ so that V is the voltage difference between the two sides of the barrier. We rewrite the current as

$$I=I_0\sin S, \tag{5.9}$$

where $S=S_1-S_2$ is the phase difference across the barrier, and the maximum supercurrent I_0 is proportional to the coupling K. Equations (5.8) and (5.9) are the governing equations for all the Josephson effects. Note that eq. (5.9) is derived under the assumption of weak coupling between the superconductors: the coupling terms in eqs. (5.1) and (5.2) are simply taken as linear in the wave functions. This is adequate for oxide barrier junctions, but in principle inadequate for more strongly coupled links. In the more general case we can write $I=I_0f(S)$, where f is periodic with period 2π. However, eq. (5.9) is generally adequate for a qualitative understanding of weak link systems, and often gives quantitative agreement. We therefore mainly restrict our attention to the simple form.

5.3   D.C. effects

When there is no voltage across the junction, eq. (5.9) gives a current determined by the phase. In practice, a given current I is passed through the junction, and for $I < I_0$ the phase S adapts so that the current passes without a voltage appearing. For $I > I_0$ a voltage V appears; the form of the I–V characteristic depends on the details of the weak link and of the circuit. A typical I–V characteristic for an evaporated junction is shown in Fig. 5.2. As the current I is increased, no voltage

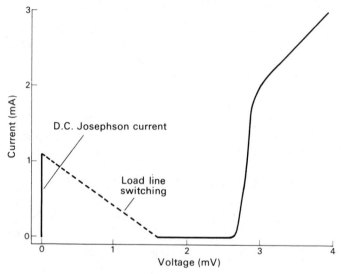

FIG. 5.2 I–V characteristic of an evaporated Pb–Pb junction at 1·2 K. (After Langenberg *et al.*, 1966.)

appears until I reaches I_0; the value of about 1 mA in Fig. 5.2 is typical. Eventually the characteristic becomes a typical SS quasiparticle tunnelling curve (compare Fig. 3.14); the intermediate behaviour is governed by the external circuitry.

The temperature dependence of I_0 for the particular case of an oxide junction was calculated from the microscopic theory by Ambegaokar and Baratoff (1963). The main features can be seen from eqs. (5.6) and (5.7), which give $I_0 \propto \sqrt{(n_1 n_2)}$. For an SS junction (identical superconductors) near to T_c we therefore have $I_0 \propto n_1 \propto T_c - T$, using the Gorter–Casimir temperature dependence. For an SS′ junction (non-identical superconductors) on the other hand, the behaviour near T_c, the lower of the two critical temperatures, is $I_0 \propto (T_c - T)^{1/2}$. For the SS junction, the theoretical result has the explicit form

$$I_0 = (\pi/2eR_n)\Delta(T)\tanh\{\Delta(T)/2k_B T\}, \tag{5.10}$$

where $\Delta(T)$ is the energy gap, and R_n is the normal state resistance of the junction. In particular, the zero-temperature result is

$$I_0(T = 0) = \pi\Delta(0)/2eR_n, \tag{5.11}$$

where $\Delta(0)$ is the energy gap at zero temperature. For a typical oxide junction 1 mm² in area, R_n is of order 1 Ω, which gives I_0 of order 1 mA (since Δ is of order 1 mV). The critical current of a poor junction falls below that given by eq. (5.11), but well-prepared oxide junctions (e.g. Schroen, 1968, Hamilton and Shapiro, 1970) can have I_0 values agreeing fairly well with that formula. Note that the current density through an oxide junction is typically only 10^3 A m⁻². This compares with, say, 10^9 A m⁻² for an irreversible type II superconductor, and 10^{11} A m⁻² for a bulk type I superconductor. The temperature dependence of I_0 given by eq. (5.10), and the corresponding temperature dependence for an SS′ junction, agree well with experimental results, as can be seen from Fig. 5.3.

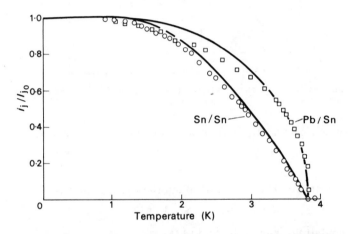

FIG. 5.3 Temperature dependence of zero-voltage current of a Sn/Sn and a Pb/Sn junction in comparison with calculation of Ambegaokar and Baratoff (1963) (solid curves). (After Fiske, 1964.)

The d.c. current is extremely sensitive to applied magnetic fields. For example, consider two identical point contacts in parallel between two bulk superconductors A and B, as sketched in Fig. 5.4. This set up is called a quantum interferometer. Suppose there is some magnetic flux Φ through the area between the junctions; this means that the vector potential **A** cannot be zero everywhere, and in particular

$$\oint \mathbf{A} \cdot d\mathbf{l} = \Phi, \tag{5.12}$$

where the integral is taken round any contour C passing through both contacts. We recall from §3.3 that a supercurrent (in the London region) is given by

$$\mathbf{J}_e = \frac{e\hbar}{m} |\psi|^2 \nabla S - \frac{2e^2}{m} |\psi|^2 \mathbf{A}. \tag{5.13}$$

The current density in the bulk of superconductors A and B is much less than that at the contacts. We have just seen that the critical supercurrent density at the

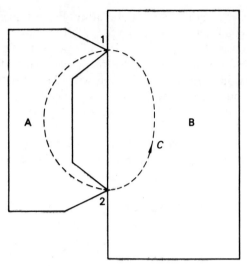

FIG. 5.4 A quantum interferometer. We label the contacts 1 and 2. C is a suitable contour for the integral in eq. (5.12).

contacts is much less than the bulk critical supercurrent. Consequently the current $\mathbf{J_e}$ in eq. (5.13) is negligible in the bulk of A and B, so we have

$$\hbar\,\nabla S = 2e\mathbf{A} \tag{5.14}$$

throughout A and B. The current through the two contacts is

$$I = I_0\{\sin(S_{1A}-S_{1B}) + \sin(S_{2A}-S_{2B})\} \tag{5.15}$$

in obvious notation. Now from eq. (5.14)

$$S_{1A}-S_{2A} = \frac{2e}{\hbar}\int_1^2 \mathbf{A}\cdot d\mathbf{l}, \tag{5.16}$$

where the integral is along any path in A from 1 to 2. Likewise

$$S_{2B}-S_{1B} = \frac{2e}{\hbar}\int_2^1 \mathbf{A}\cdot d\mathbf{l}. \tag{5.17}$$

We therefore have

$$(S_{1A}-S_{2A}) + (S_{2B}-S_{1B}) = \frac{2e}{\hbar}\oint \mathbf{A}\cdot d\mathbf{l} = 2\pi\Phi/\phi_0, \tag{5.18}$$

using eq. (5.12) for the flux Φ between the contacts. For simplicity we shall assume that Φ is just the flux produced by an external applied field. In general Φ contains a part LI_s produced by any current I_s circulating in the loop (L is the self inductance of the loop), but this is negligible if $LI_0 \ll \phi_0$, where I_0 is the critical current of either junction.

The significance of eq. (5.18) is seen if we reorganize eq. (5.15):

$$I = I_{max}\sin S \tag{5.19}$$

with

$$I_{max} = 2I_0\cos(\pi\Phi/\phi_0) \tag{5.20}$$

and $S = \frac{1}{2}(S_{1A} + S_{2A} - S_{1B} - S_{2B}).$ (5.21)

We have used eq. (5.18) to express the argument of the cosine function in terms of the magnetic flux Φ, rather than the sum of phases occurring on the left-hand side of that equation. We read eq. (5.19) as we read the original eq. (5.9). For an applied current $I < I_{max}$, S adjusts so that the current is passed without a voltage appearing. However, in the present case the maximum current I_{max} is modulated by the magnetic flux trapped between the junctions. The period of modulation is $\phi_0 = 2 \times 10^{-15}\ \text{Wb} = 2 \times 10^{-7}\ \text{G cm}^2$.

Equation (5.15) shows clearly why the arrangement we have been describing is called an interferometer; the total current is the coherent sum of the separate currents through each junction. Typical results for the current through an interferometer are shown in Fig. 5.5.

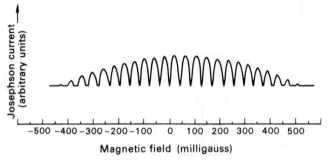

FIG. 5.5   Experimental trace of I_{max} versus magnetic field for a two-junction interferometer. The envelope arises because of the finite size of the individual junctions. (After Jaklevic et al., 1965.)

A thin film junction is similar to the interferometer we have just described, in that the maximum supercurrent through the junction depends on the magnetic field in the plane of the oxide. As shown in Fig. 5.6, the field penetrates through a total area $(\lambda_A + \lambda_B + d)L$, where λ_A, λ_B are the penetration depths, d the oxide thickness, and L the length of the junction. A calculation similar to the one we have just given (the sum over two junctions is replaced by an integral along the oxide) yields

$$I_{max} = I_0 \frac{\sin (\pi\Phi/\phi_0)}{\pi\Phi/\phi_0},$$ (5.22)

where $\Phi = BL(\lambda_A + \lambda_B + d)$ is the flux trapped in the junction. Equation (5.22) is of course just the form derived in optics for the Fraunhöfer diffraction amplitude of a single slit. Good experimental agreement can be found with eq. (5.22), as shown in Fig. 5.7.

We have seen that in both the two-junction interferometer and the evaporated junction, the critical current oscillates with period ϕ_0 in the applied magnetic flux. In the interferometer, in particular, the critical current oscillates very rapidly with applied field if the area between the junctions is made sufficiently large. As might be expected, interferometers have therefore been applied as extremely sensitive magnetometers. In fact, magnetometers are more often made with a supercon-

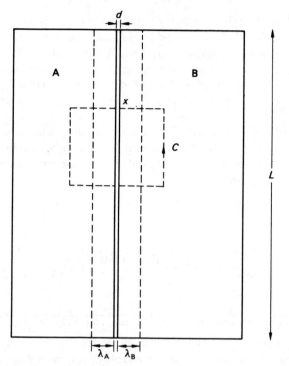

FIG. 5.6   Section through thin-film junction between A and B.

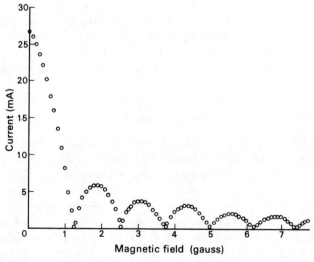

FIG. 5.7   Magnetic-field dependence of the maximum d.c. Josephson current of a Sn–Sn junction at 1·2 K. (After Langenberg *et al.*, 1966.)

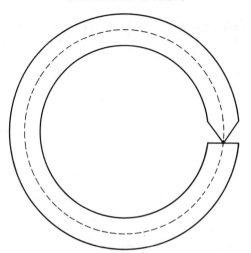

FIG. 5.8 Superconducting loop with a single weak link. The contour C used for integration is shown by the broken line.

ducting loop containing a single weak link, as sketched in Fig. 5.8. As we shall see, the current through the link is an oscillating function of the flux applied to the loop, so that the one-link loop retains the magnetic field sensitivity of the interferometer. Both one- and two-junction loops are called SQUIDS (superconductive quantum interference detectors), and because of their increasing importance in instrumentation, as well as their intrinsic interest, we shall now discuss the general behaviour in an applied field of a loop containing one weak link. The original work on loops of this kind was done by Silver and Zimmerman and their co-workers, and they also developed the techniques for viable magnetometers. Our discussion will follow part of the basic paper by Silver and Zimmerman (1967).

We start, naturally, with the London equation for the current, eq. (3.18), which we integrate round the contour C shown dotted in Fig. 5.8. The integral of $\mathbf{A}$ gives the magnetic flux through the ring, as usual:

$$\oint \mathbf{A} \cdot d\mathbf{l} = \Phi. \tag{5.23}$$

It is important now to include the magnetic field generated by the current I circulating in the ring; the corresponding flux is LI, where L is the self-inductance of the ring. We therefore write

$$\Phi = \Phi_x + LI, \tag{5.24}$$

where Φ_x is the applied flux, generated for example by a coil placed within the ring. Φ_x is what one sets out to measure in a magnetometer. The integral of ∇S round C gives $2n\pi$ as usual. The integral of $\mathbf{J}_e$ splits into two parts. In the bulk of the superconductor $\mathbf{J}_e$ is zero, so the integral is zero. Across the junction, S is changing rapidly, so we have

$$\int \mathbf{J}_e \cdot d\mathbf{l} = \frac{e\hbar}{m} |\psi|^2 \int \nabla S \cdot d\mathbf{l} = \frac{e\hbar}{m} |\psi|^2 S, \tag{5.25}$$

where the integrals are across the junction, and S is the total phase difference across the junction. We neglect the integral of $\mathbf{A}$ in eq. (5.25), since it is proportional to the length of the junction in the direction of the current, and therefore very small. Putting together eqs. (5.23) and (5.25), we find

$$S + 2\pi\Phi/\phi_0 = 2n\pi. \qquad (5.26)$$

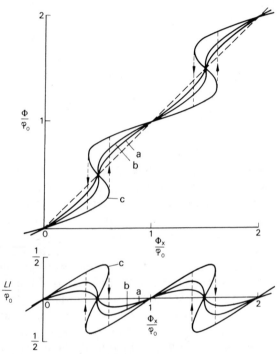

FIG. 5.9   Magnetic flux Φ and circulating current I as functions of external flux Φ_x of a superconducting ring incorporating one Josephson tunnelling junction. The three sets of curves correspond to (a) $LI_0/\phi_0 = 1/4\pi$; (b) $LI_0/\phi_0 = 1/2\pi$; (c) $LI_0/\phi_0 = 1/\pi$. (After Silver and Zimmerman, 1967.)

Finally, the supercurrent I flowing round the loop is given by the usual Josephson equation:

$$I = I_0 \sin S. \qquad (5.27)$$

Equations (5.24), (5.26) and (5.27) are linked equations for the three unknowns Φ, I and S in terms of the applied flux Φ_x. We are particularly interested in the behaviour of Φ and I as functions of Φ_x. The form of this behaviour depends on the dimensionless parameter LI_0/ϕ_0; the details of the calculation are left for a problem, and results for three values of LI_0/ϕ_0 are shown in Fig. 5.9. For $LI_0/\phi_0 < 1/2\pi$, Φ is a single-valued function of Φ_x, whereas for $LI_0/\phi_0 > 1/2\pi$, Φ is three valued for some parts of the range of Φ_x. In the latter case, hysteresis can occur, with the transitions in increasing and decreasing field occurring at different Φ_x-values, as shown by the dotted lines in Fig. 5.9. The limiting forms of behaviour are $\Phi = \Phi_x$ for $LI_0 = 0$, which corresponds to an open ring, and complete flux quantization $\Phi = n\phi_0$ for $LI_0 \gg \phi_0$, which corresponds to a closed ring with no weak link. Curves like those in Fig. 5.9 represent intermediate forms of behaviour between these two extremes. Experimental results for the static magnetization of a weakly connected ring are shown in Fig. 5.10.

The application of a weakly connected ring as a magnetometer requires some means of sensing the position of the ring on one of the response curves of Fig. 5.9. This is frequently done by looking at the r.f. response of an LC circuit inductively coupled to the ring. In addition, a practical device must be sufficiently robust; in particular it is desirable that it should be adjusted once and for all, and not need adjustment every time it is used. A suitable arrangement was first described by Zimmerman et al. (1970), and is shown in Fig. 5.11. For flexibility of design, the flux to be measured is usually coupled into the weakly connected ring, cavity 2, by means of the *flux transformer* C_1C_2. This consists of a completely superconducting wire with the flux to be measured enclosed by the loop C_2; since the total flux enclosed by the wire is constant, the flux through C_1 decreases as the flux through C_2 increases, and vice versa. The point contact is adjusted so as to have a critical current I_0 giving LI_0/ϕ_0 somewhat larger than $1/2\pi$, so that there is some loss when the flux Φ_x in cavity 2 is cycled over a range of values including one or more of the hysteresis loops in Fig. 5.9. When the r.f. current $I_{r.f.}$ is such that the critical current of the point contact is exceeded at some point of the r.f. cycle, flux moves to and fro between cavities 1 and 2 during the cycle. The resulting dissipation appears as a voltage $V_{r.f.}$, whose magnitude depends on the flux Φ_x already in cavity 2. Fig. 5.12(a) shows characteristics $V_{r.f.}$ versus $I_{r.f.}$ for two different values of Φ_x, and Fig. 5.12(b) shows $V_{r.f.}$ versus Φ_x for four different fixed values of $I_{r.f.}$. In both cases, of course, the response is periodic in Φ_x with period ϕ_0.

There are various ways of using the characteristics of Fig. 5.12 to measure the flux Φ_x. For relatively coarse measurements, the number of zero crossings in $V_{r.f.}$ as Φ_x is increased can be counted to give a digital read-out of the number of flux quanta in Φ_x (Forgacs and Warnick, 1967). For measurements of the departure of Φ_x from the nearest whole number of quanta $n\phi_0$, Φ_x can be modulated with amplitude ϕ_0 at some audio frequency. The magnitude of the signal $V_{r.f.}$ detected at the audio frequency then depends on the difference $\Phi_x - n\phi_0$. Further details of the operation of SQUIDS as magnetometers and voltmeters, and the noise limitations on their sensitivity, are given in various articles and reviews (Zimmerman et al., 1970, Zimmerman, 1971a, b, Webb, 1972).

The results we have obtained so far mainly concern the supercurrent at strictly zero voltage. Finally, in this section, we turn to the influence of circuit parameters on the d.c. $I-V$ characteristics of a single junction near $V = 0$. There are two

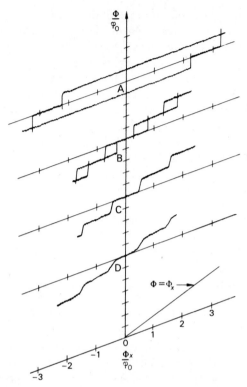

FIG. 5.10 Magnetic behaviour of a weakly connected ring with IL/ϕ_0 equal to $3\frac{1}{4}$, $\frac{3}{4}$, $\frac{1}{2}$ and $\frac{2}{5}$ respectively for curves A, B, C, D. (After Silver and Zimmerman, 1967.)

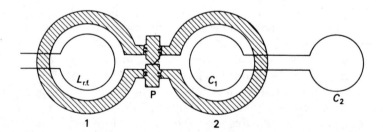

FIG. 5.11 A symmetric SQUID. The point contact P is made between two niobium screws, and is symmetrically placed between two cavities 1 and 2 in niobium (shaded). Cavity 1 contains the coil of the RF circuit. Cavity 2, which is the weakly connected ring, is linked to the flux to be measured by the flux transformer $C_1 C_2$, which is made of superconducting wire. (After Zimmerman *et al.*, 1970.)

factors involved in general: the nature of the external circuit driving the junction, and the form of the *background current* which the junction would carry in the absence of a supercurrent. The background current of an oxide junction is the quasiparticle tunnelling characteristic, as in Fig. 5.2, and the background current of a point contact junction is generally taken to be Ohmic; in any case we write

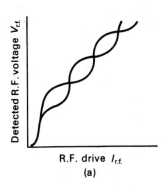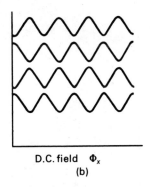

FIG. 5.12   (a) $V_{r.f.}$ as a function of $I_{r.f.}$ for $\Phi_x = n\phi_0$ and $\Phi_x = (n+\frac{1}{2})\phi_0$. (b) $V_{r.f.}$ as a function of Φ_x for four different values of $I_{r.f.}$. (After Zimmerman et al., 1970.)

the background current as $I_n(V)$. We assume that this current is in parallel with the Josephson current, so that the governing equations are

$$I = I_0 \sin S + I_n(V),$$ (5.28)

$$\hbar \frac{dS}{dt} = 2eV.$$ (5.29)

We consider explicitly just two extreme forms of external circuit conditions, voltage feed and current feed. With voltage feed, Fig. 5.13(a), V is fixed and we determine I, whereas with current feed, Fig. 5.13(b), I is fixed and we determine V.

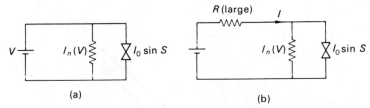

FIG. 5.13   Circuits for (a) voltage feed and (b) current feed.

Unless special care is taken, experiments generally use current feed, since the d.c. impedance of the junction is low. We shall also simplify the problem by taking the background current as Ohmic:

$$I_n(V) = G_n V.$$ (5.30)

For voltage feed, eq. (5.29) gives

$$S = S_0 + 2eVt/\hbar,$$ (5.31)

and the time average of eq. (5.28) gives for the d.c. current

$$I_{d.c.} = I_0 \sin S \qquad \text{for } V = 0$$ (5.32)

$$= I_n(V) \qquad \text{for } V \neq 0.$$ (5.33)

The Josephson current is a zero-voltage spike, as in Fig. 5.14(a). For current feed,

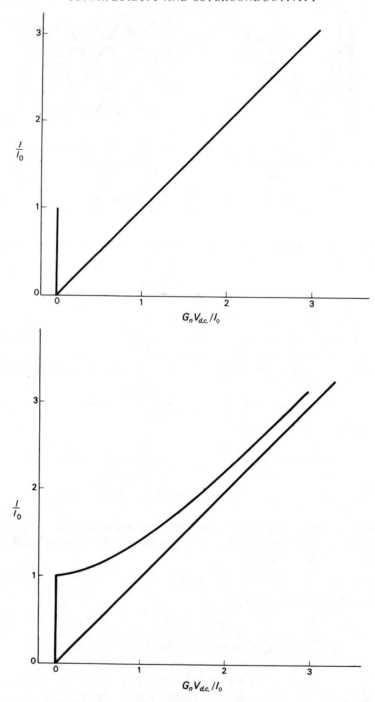

F$_{IG}$. 5.14   Current–voltage characteristics. (a) Voltage feed, corresponding to F$_{IG}$. 5.13(a) and eqs. (5.32) and (5.33). (b) Current feed, corresponding to Fig. 5.13(b) and eq. (5.35).

on the other hand, S is no longer linear in time, and consequently $I_0 \sin S$ does not average to zero. Since I is fixed now, and not V, we substitute for V in eq. (5.28), using eq. (5.29):

$$\frac{\hbar}{2e} G_n \frac{dS}{dt} + I_0 \sin S = I, \qquad (5.34)$$

where we have taken $I_n(V)$ to be Ohmic. Equation (5.34) can be solved for $S(t)$ with the help of standard tables of integrals. Once $S(t)$ is known, then the d.c. voltage $V_{d.c.}$ across the junction can be found by means of eq. (5.29) from the time average of dS/dt; the result is (McCumber, 1968, Stewart, 1968)

$$(G_n V_{d.c.})^2 + I_0^2 = I^2. \qquad (5.35)$$

This solution is shown in Fig. 5.14(b); as I increases, $V_{d.c.}$ gradually approaches the background value. We may say that for $I > I_0$ the phase S precesses in such a way as to spend more time on values for which the supercurrent is positive.

The influence of a specific circuit can always, in principle, be dealt with by the technique we have outlined. Obviously a considerable variety of behaviour is possible. The difference in d.c. characteristics between oxide and point contact junctions can be understood, at least partially, in this way. The oxide junction has a large capacitance in parallel, and the point contact a large inductance in series. McCumber (1968) found the d.c. characteristics for these two equivalent circuits (with current and voltage feed respectively); the results, which are in qualitative agreement with experimental characteristics, are shown in Figs. 5.15 and 5.16. In

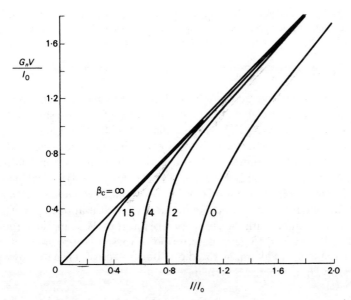

FIG. 5.15 Current–voltage characteristic for a current-fed junction with a capacitance in parallel. Note that current is plotted along the horizontal axis. β_c is the capacitance in reduced units: $\beta_c = 2eI_0 C/\hbar G^2$. (After McCumber, 1968.)

addition, the capacitive loading leads to a hysteresis in the characteristic (Stewart, 1968); experimental measurements (Scott, 1970) are in good agreement with the theoretical predictions. An extensive review of equivalent circuit calculations has been given by de Bruyn Ouboter and de Waele (1971).

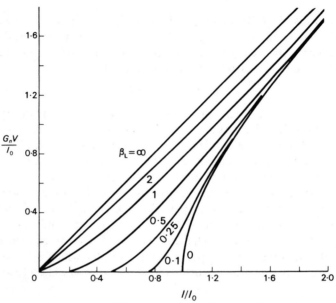

FIG. 5.16   Current–voltage characteristics for a voltage-fed junction with an inductance in series. β_L is the inductance in reduced units: $\beta_L = \hbar/2eI_0 L$. (After McCumber, 1968.)

5.4   A.C. effects

When there is a voltage across a weak link, the phase difference between the sides precesses at the rate given by the usual relation for phase slip, eq. (5.8). The simplest example is a weak link voltage biassed at V. There is then an energy difference $\mu_1 - \mu_2 = 2\,eV$ between the Cooper pairs on either side, and Cooper pairs can continue to tunnel with the emission of a photon of frequency $\omega_0 = 2\,eV/\hbar$. The conversion factor gives 484 MHz for 1 μV, so typical biases of the order of a few millivolts correspond to microwave photons. Similarly, if a junction is biassed at V and a microwave field of frequency $2\,eV/\hbar$ is applied to the junction, Cooper pairs can tunnel with the absorption of photons. We can use the governing equations (5.8) and (5.9) to get a description of these two processes.

Suppose we have a weak link with a d.c. voltage V across it in a microwave field. The phase-slip equation, (5.8), is then

$$\hbar \frac{dS}{dt} = 2\,eV + 2\,ev \cos{(\omega t + \theta)}. \qquad (5.36)$$

Note that we are here assuming for simplicity that the weak link is voltage fed. We show the phase of the microwave field, θ, explicitly. Equation (5.36) gives for the superfluid phase

$$S = \frac{2\,eV}{h}t + \frac{2\,ev}{h\omega}\sin(\omega t + \theta) + S_0, \tag{5.37}$$

and this is to be substituted in the current $I_0 \sin S$. The resulting expression can be analysed with the aid of the identity

$$\exp(iC \sin x) = \sum_{n=-\infty}^{\infty} J_n(C)\exp(inx). \tag{5.38}$$

Here $J_n(C)$ is the Bessel coefficient of order n. We find for the current across the weak link

$$I = I_0 \sum_{n=-\infty}^{\infty} (-1)^n J_n\left(\frac{2\,ev}{h\omega}\right)\sin\left\{\left(\frac{2\,eV}{h} - n\omega\right)t - n\theta + S_0\right\}. \tag{5.39}$$

The most important feature of this expression is that there is a d.c. supercurrent when

$$2\,eV = n\hbar\omega. \tag{5.40}$$

The magnitude is

$$I_n = (-1)^n I_0 J_n\left(\frac{2\,ev}{h\omega}\right)\sin(S_0 - n\theta). \tag{5.41}$$

This is like the zero-voltage supercurrent, in that the phase S_0 can adjust to match the applied d.c. current I_n over a certain range. We have a *constant voltage step*, of magnitude

$$I_n^{(\text{step})} = 2I_0 J_n\left(\frac{2\,ev}{h\omega}\right). \tag{5.42}$$

In addition, the zero-voltage supercurrent is depressed:

$$I_0^{(\text{step})} = 2I_0 J_0\left(\frac{2\,ev}{h\omega}\right). \tag{5.43}$$

For $2\,ev/h\omega \ll 1$, we can expand the Bessel coefficient here, to get

$$I_0^{(\text{step})} = 2I_0(1 - 2\,e^2v^2/h^2\omega^2). \tag{5.44}$$

Equation (5.40), with $n = 1$, is exactly the condition for tunnelling with absorption of one photon, that we described earlier; eq. (5.42) gives the corresponding supercurrent step. For general n, eq. (5.40) is the condition for tunnelling with absorption of n photons simultaneously; again this gives a supercurrent step.

Constant voltage steps in an applied microwave field were first observed by Shapiro (1963); typical more recent results are shown in Fig. 5.17. It can be seen from there that the step amplitudes oscillate as a function of microwave intensity, qualitatively in agreement with the Bessel function dependence of eq. (5.42). For example, as the power is increased, the $n = 2$ step rises from zero to a maximum, then decreases, and so on. It is usually found that quantitative agreement with the Bessel function dependence is fairly good for point-contact junctions, except that

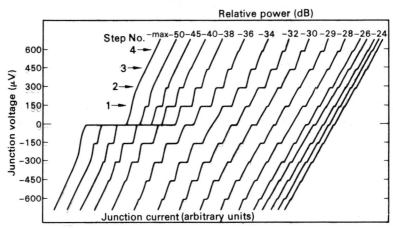

FIG. 5.17   Voltage–current curves for a Nb–Nb point-contact junction at 4·2 K exposed to a 72 GHz signal at various power levels. (After Grimes and Shapiro, 1968.)

the zero-voltage step is excessively large at low powers. The reason for this excess is not altogether clear; the explanation given by Grimes and Shapiro (1968) has been criticized by Zimmerman (1970).

We mentioned at the end of § 5.2 that in a microbridge (Fig. 5.1), for example, the current should be taken simply as a periodic function $I_0 f(S)$ of the phase. Although the difference between $f(S)$ and a simple sine wave is often unimportant, it does make a significant difference to the driven a.c. effect we are discussing. We can expand $f(S)$ as a Fourier sum of sine waves:

$$f(S) = \sum_{n'} A_{n'} \sin n'S. \tag{5.45}$$

With this expression for the current, we can generalize the analysis which led to the synchronization condition of eq. (5.40) (see Problem 5.7). Equation (5.39) is replaced by an expression with an additional sum over n', and the synchronization condition becomes

$$2eV = n\hbar\omega/n'. \tag{5.46}$$

Experimental results for the driven a.c. effect in a microbridge are shown in Fig. 5.18. The subharmonic series with $n = 1$ is well defined down to $n' = 6$ and in addition steps are observed at positions in which both n and n' are relatively simple integers different from unity.

It is worth remarking that because we assumed voltage feed, that is, a high specimen impedance to both d.c. and microwaves, eq. (5.42) predicts, strictly speaking, a current spike at exactly the synchronization voltage $n\hbar\omega/2e$. However, the experimental results show a step with a gradual return to the background current; by analogy with the zero-voltage current we can see that this is indeed what is to be expected with a current-fed specimen. As for the zero-voltage current, if we assumed current feed we should get different, and more complicated, governing equations for the microwave-induced steps. The appropriate equations have been written down by Waldram *et al.* (1970), but analytic solutions do not appear to be possible.

At first sight the a.c. effect might appear to promise a useful microwave source. However, the low impedance of all weak links, of order $1\,\Omega$ or less, means that

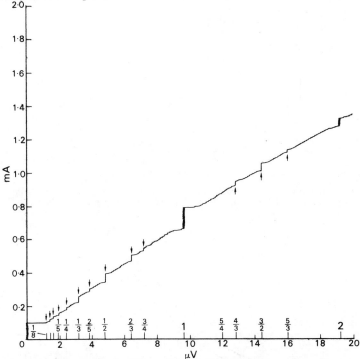

FIG. 5.18 Current–voltage characteristic of a thin-film microbridge in a microwave field of frequency 4·26 GHz. (After Dayem and Wiegand, 1967.)

there is always a large impedance mismatch between the weak link and any external circuitry. Consequently external power levels are always very low; emitted power has been detected, but is generally only of order 10^{-11} W (Yanson *et al.*, 1965, Langenberg *et al.*, 1965). In fact, because of the impedance mismatch, a high Q standing-wave pattern is set up at resonant frequencies of an oxide junction, which may be regarded as a short piece of transmission line. If a magnetic field is applied in the plane of the junction, the Josephson current can be modulated spatially to match the standing wave. When this happens, a high microwave intensity is generated in the junction at a d.c. bias voltage $V_c = n\hbar\omega_c/2e$, where ω_c is the resonant frequency; the coupling back of the microwave field then induces a constant-voltage current step (Fiske, 1964, Coon and Fiske, 1965). An I–V characteristic displaying these Fiske steps is shown in Fig. 5.19.

The a.c. effect has found useful applications in which an external microwave field is applied to a weak link. The most important is the accurate measurement of

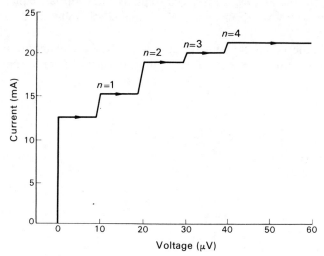

FIG. 5.19 I–V characteristic for a Sn–Sn thin film junction at $T = 1\cdot2$ K with a magnetic field of $1\cdot9$ gauss applied in the plane of the film. (After Langenberg *et al.*, 1966.)

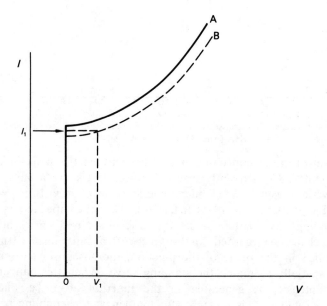

FIG. 5.20   To illustrate the broad-band detection method of Grimes *et al.* (1966). Curve A: characteristic in the absence of applied radiation. Curve B: characteristic in the presence of radiation. The junction is biassed at current I_1, and a voltage V_1 is developed when radiation is applied.

the fundamental constant h/e. As we have seen, a microwave field of frequency ω induces a constant voltage ($n = 1$) step at voltage V, with $V/\omega = \hbar/2e$. The Josephson effect therefore measures h/e, or equivalently gives an expression for a voltage standard in terms of a frequency standard. The original measurement of h/e (Parker et al., 1967) gave a significant revision of the accepted value, and removed an apparent discrepancy between calculations based on quantum electrodynamics and the energy-level structure of hydrogen. Later measurements led to a revision of the values of all the fundamental constants; a full account is given by Taylor et al. (1969). The limitation on the accuracy to which h/e can be determined arises because the voltage at the junction is of the order of a few millivolts, whereas voltage standards produce about 1 V. Consequently elaborate voltage dividers have to be used, and these are the dominant source of error. Currently, the accuracy is better than 1 in 10^6 (Finnegan et al., 1970).

Weak links can be used as microwave and far infrared detectors. We saw (eqs. 5.43 and 5.44) that the zero-voltage current step is depressed by incident radiation of any frequency. Grimes et al. (1966) used this effect in a broad-band detector. The weak link (Grimes et al. used various point contacts) is current biassed to a point near the top of the zero-voltage step. In the presence of radiation, the step is diminished, and a voltage appears across the junction—see Fig. 5.20. In practice, the incident radiation is chopped at some frequency, and the voltage at that

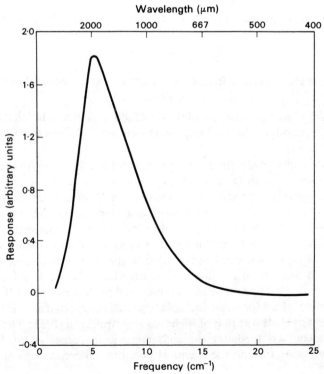

FIG. 5.21 Spectral response of a typical In–In point contact at 2·2 K. (After Grimes et al., 1966.)

frequency is monitored. As can be seen in Fig. 5.21, the spectral response of an In–In contact at 2·2 K peaks in the sub-millimetre region, and is in fact somewhat better than that of any other detector in that region.

5.5   The search for Josephson effects in liquid helium

We have seen how the use of a macroscopic wave function to describe a superconductor leads to the prediction of the various Josephson effects which are observed when two bulk superconductors are weakly coupled together. Naturally, the description of superfluid helium in the same terms raises the expectation of analogous effects when two volumes of bulk helium are connected by a weak link, that is a region in which the wave functions appropriate to each reservoir can overlap. We consider the situation (Fig. 5.22) in which there is a level difference ΔZ between the two reservoirs; for the moment we assume that the liquid is at the same temperature in both. There is, therefore, a chemical potential difference

$$\Delta\mu = \mu_2 - \mu_1 = m_4 g \Delta Z. \tag{5.47}$$

In § 4.5 we introduced the idea of phase slip by means of the basic equation we first met in Chapter 2 (eq. (2.16)). Its relevance for the Josephson effect has already been pointed out in § 5.2, and we now use it again,

$$\frac{d}{dt}(S_2 - S_1) = -\frac{1}{\hbar}(\mu_2 - \mu_1). \tag{5.48}$$

Here S_1 is the phase of reservoir 1, and S_2 that of reservoir 2. By writing

$$\frac{d}{dt}(S_2 - S_1) = -2\pi\nu \tag{5.49}$$

we can define a characteristic frequency for phase slip at the weak link,

$$\nu = m_4 g \Delta Z / h \tag{5.50}$$

where we have used eqs. (5.47), (5.48) and (5.49) together. The characteristic frequency corresponds to the a.c. Josephson frequency $2eV/h$ for the tunnelling of Cooper pairs (§ 5.4).

The main difficulty in devising experiments to look for Josephson effects in He II is the design of a suitable weak link between the reservoirs. It has been suggested that quantum mechanical tunnelling of helium atoms could occur through a barrier pierced with holes of atomic dimensions (Mamaladze and Cheishvili, 1966), but such a junction would not be simple to construct. There is also the possibility that a helium film might serve as a weak link (Donnelly, 1965), but no film flow effect that can be attributed to quantum interference has been reported yet. In fact, in all the known investigations, the weak link has been provided by a hole 10–20 μm in diameter, punched in thin metal foil (Fig. 5.23).

The experiment which for a considerable time was interpreted as evidence for a Josephson effect in He II was that of Richards and Anderson (1965). The purpose of the experiment was to detect the characteristic frequency of eq. (5.50) by imposing an a.c. variation on the chemical potential difference at another fixed frequency v_0, and looking for synchronization of the two according to the relation

$$n_1 v_0 = n_2 v \tag{5.51}$$

where n_1 and n_2 are small integers. Equations (5.50) and (5.51) together give

$$(\Delta Z)_j = \frac{n_1}{n_2} \frac{h v_0}{m_4 g} \tag{5.52}$$

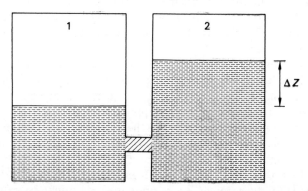

FIG. 5.22 Two reservoirs of bulk liquid helium connected by a weak link. Level difference ΔZ gives chemical potential difference $m_4 g \Delta Z$ when reservoirs are equal in temperature.

from which it is to be expected that the system should show anomalies in the superfluid flow through the hole at level differences which are multiples or sub-multiples of $h v_0/m_4 g$.

The experimental arrangement is shown schematically in Fig. 5.23: one reser-

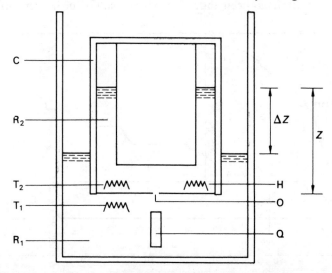

FIG. 5.23 Schematic of apparatus used by Richards and Anderson (1965) to look for a.c. Josephson effect in He II. R_1, R_2 —reservoirs; O —orifice; Q —transducer; C —annular capacitor for monitoring level of R_2 (capacitor for R_1 not shown). Modifications introduced by Musinski and Douglass (1972): H —heater; T_1, T_2 —resistance thermometers.

voir is cylindrical and is connected to the outer reservoir through the orifice in its base. The liquid levels are monitored by a capacitative technique. A quartz crystal transducer placed just below the orifice is used to irradiate the weak link with a sound field of frequency v_0. The decay of the level difference by gravitational flow through the orifice was observed: with the transducer not operating, the curve of ΔZ against time was smooth, but when the transducer was switched on, the curve showed a step-like structure, indicating that the system had a preference for sticking at certain level differences for some time, during which there was no net flow through the orifice.

The step structure was more clearly seen when beginning from $\Delta Z = 0$. In this case, the transducer acted as a fluid pump and increased the level difference with time. Figure 5.24 shows results from a subsequent repeat of the experiment (Richards, 1970). The lowest curve was interpreted as indicating a marked stability at a level difference equal to $hv_0/m_4 g$. When the transducer was operated at higher power (centre curve), steps at higher multiples with n_1/n_2 varying from 2 to 6 and also some submultiples at fractions of $\frac{1}{2}$, $\frac{1}{3}$ and $\frac{1}{4}$ of the basic step were seen. The curve obtained with the maximum power from the transducer also shows steps, although their stability is much less pronounced, as might be expected. The experiment was also repeated by Khorana (1969) and Hulin et al. (1972) with essentially the same results. The stable situations observed in this way were regarded as analogous to the current steps seen, for example, in the superconducting microbridge experiment of Dayem and Wiegand (1967) shown earlier in Fig. 5.18. The Richards–Anderson steps were therefore interpreted as evidence for the occurrence of the driven a.c. Josephson effect in He II.

This interpretation has been undermined as a result of further repetitions of the

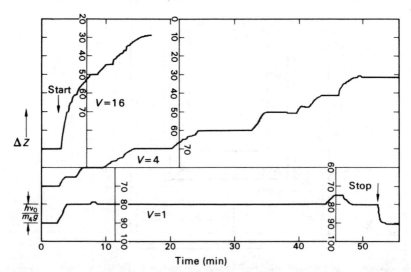

FIG. 5.24   Measured values of head difference ΔZ versus time for various values of a.c. voltage drive (V in arbitrary units) on the transducer (104·6 kHz). Diameter of orifice 12 μm. Horizontal regions indicate dynamic stability of the He at finite ΔZ. (After Richards, 1970.)

Richards–Anderson experiment by Musinski and Douglass (1972) and by Leiderer and Pobell (1973). In each case the basic experiment was extended by allowing additional variation of the experimental parameters. Musinski and Douglass placed a heater in one of the reservoirs and measured the temperature difference across the orifice (Fig. 5.23); they also used a transducer with variable frequency. In earlier versions of the experiment care was taken to keep the two reservoirs at the same temperature: when there is a difference $\Delta T = T_2 - T_1$, then the chemical potential difference becomes

$$\Delta \mu = \mu_2 - \mu_1 = m_4 g \Delta Z - m_4 \sigma \Delta T \tag{5.53}$$

where σ is the entropy per unit mass. Using eq. (5.53) instead of eq. (5.46) but otherwise the same argument which led to eq. (5.52), we see that Josephson steps in the ΔZ versus time curve are expected at level differences

$$(\Delta Z)_J = \frac{n_1}{n_2} \frac{h v_0}{m_4 g} + \frac{\sigma}{g} \Delta T. \tag{5.54}$$

To test this argument Musinski and Douglass suggested three criteria which must be satisfied by any steps that are observed. (A) A step once established with no temperature difference should respond in such a way that $(\Delta Z)_J$ is proportional to ΔT when a temperature difference is subsequently applied (eq. (5.54)). (B) Stable states of the system should occur at particular values of the level difference ΔZ, and should be independent of Z, the height of the liquid in the inner reservoir (Fig. 5.23). (C) When the sound field frequency v_0 is varied, the value of $(\Delta Z)_J$ should respond according to

$$\frac{d(\Delta Z)_J}{dv_0} = \frac{n_1}{n_2} \frac{h}{m_4 g} \tag{5.55}$$

which follows from eq. (5.54).

By monitoring the level difference versus time with the transducer operating at a fixed frequency, Musinski and Douglass found many steps of the Richards–Anderson type (e.g. Fig. 5.25). However, they observed that the values of ΔZ corresponding to steps changed very little when a temperature bias ΔT was applied, thus failing criterion A.

The curves in Fig. 5.25 show the results of two runs taken under identical conditions, except that the position corresponding to $\Delta Z = 0$ was changed by raising the level of the outer bath. Treated as ΔZ versus time plots, the curves are quite unlike each other. On the other hand, when the variation in the height Z is taken into account and the curves are replotted as in Fig. 5.25, there is a remarkable similarity between them. These results suggest that steps occur at particular values of Z and not of ΔZ, a conclusion which is strengthened by monitoring ΔT at the same time as ΔZ. Here the heater is switched off and variations in ΔT across the orifice are attributed to the action of the transducer. The temperature curves corresponding to the two runs in Fig. 5.25 are drawn as ΔT versus Z plots in Fig. 5.26 and again there is a close similarity between the two. Thus the steps fail criterion B.

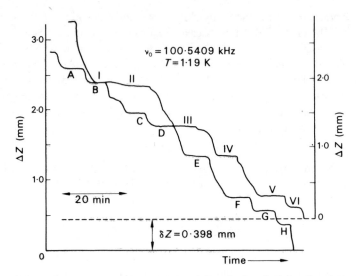

FIG. 5.25 Curves showing variation of ΔZ with time for gravitationally driven flow through the weak link with transducer operating. Conditions for two runs were identical except that level in R_1 was raised for second curve. When this difference is allowed for there is apparent correspondence between steps B and I, D and III, E and IV, F and V and G and VI. (Musinski and Douglass, 1972.)

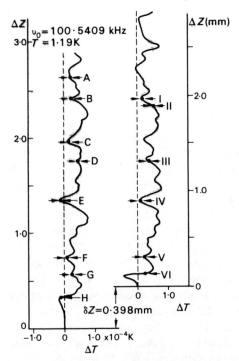

FIG. 5.26 Curves corresponding to runs in Fig. 5.25 showing variation of ΔZ versus ΔT. When initial difference in $\Delta Z = 0$ point is again allowed, curves show clear similarity. Labels indicate position of steps in Fig. 5.25. (Musinski and Douglass, 1972.)

Finally, when the applied frequency ν_0 was varied with all other parameters constant, flow curves taken over the same region of the capacitor indicated that the step height decreased with frequency, whereas eq. (5.55) predicts an increase. Therefore, criterion C is failed as well. The inevitable conclusion drawn by Musinski and Douglass is that the Richards–Anderson steps are not Josephson steps.

The work of Leiderer and Pobell (1973) also leads to this conclusion and suggests what may be the true nature of the Richards–Anderson steps. In the same way as before, with the transducer operating at 102 kHz, Leiderer and Pobell found stable states at certain values of the liquid level Z (Fig. 5.23), very similar to those shown in Fig. 5.25. The experiment was then performed again with a transducer operating at a very different frequency, 168 kHz. Again a step-shaped profile for Z with time was found. According to eq. (5.55), the step-height for Josephson steps should increase linearly with the transducer frequency ν_0. In fact, it was found that the observed step-height decreased and was proportional to ν_0^{-1}. Furthermore, at both frequencies, the step-height was close to one half-wavelength of the ultrasound in He II. Consequently, Leiderer and Pobell suggested that the stability of the liquid level at particular values of Z was associated with standing sound waves in the reservoir above the orifice and not with the Josephson effect.

If the Richards–Anderson experiment has been interpreted wrongly, this may be because of an unfortunate near-coincidence. The original experiment, and those repetitions which supported it, used transducers operating very near to 100 kHz. It so happens that, when $\nu_0 = 100$ kHz, the basic Josephson step-height $h\nu_0/m_4 g$ is equal to 1·02 mm, whilst the half-wavelength of ultrasound $u_1/2\nu_0$ is 1·19 mm when the temperature is 1·2 K.

It may be thought disappointing that no undisputed observation of Josephson effects in He II has yet been made. The results of future attempts at detection will be matters of keen interest.

Problems

5.1 Check that eqs. (5.6) to (5.9) do follow from the previous equations, as stated.

5.2 Derive eq. (5.22) for the field dependence of the maximum supercurrent through a thin film junction.

(*Hint*: Let $S(x)$ be the phase difference between A and B at distance x along the barrier—see Fig. 5.6. Take the origin of x in the centre of the junction, so that the junction extends from $-L/2$ to $L/2$. Show by an appropriate integration round the contour C that

$$S(x) - S(0) = \frac{2e}{\hbar} B(\lambda_A + \lambda_B + d)x.$$

The total supercurrent through the junction is

$$I = \frac{I_0}{L} \int_{-L/2}^{L/2} \sin S(x)\,dx,$$

where we write the constant as I_0/L, so that I_0 is the maximum supercurrent in the absence of a magnetic field. After some manipulation, the integral gives eq. (5.22).)

5.3 Combine eqs. (5.24), (5.26) and (5.27) to obtain the following implicit equations for the flux Φ and the circulating current I in a one junction SQUID:

$$I = I_0 \sin(2n\pi - 2\pi\Phi_x/\phi_0 - 2\pi LI/\phi_0)$$

$$\Phi - \Phi_x = LI_0 \sin(2n\pi - 2\pi\Phi/\phi_0).$$

Show qualitatively that the equation for Φ does lead to the kind of dependence shown in Fig. 5.9. (It is easier to consider Φ_x as a function of Φ, rather than vice versa).

5.4 Tables of integrals (e.g. Dwight, 1961) give

$$\int \frac{dx}{a+b\sin x} = \frac{2}{(a^2-b^2)^{1/2}} \tan^{-1}\left\{\frac{a\tan(x/2)+b}{(a^2-b^2)^{1/2}}\right\}$$

for $a^2 > b^2$.

Use this form to integrate eq. (5.34), with the result

$$I\tan(S/2) = I_0 + (I^2 - I_0^2)^{1/2} \tan\{e(I^2 - I_0^2)^{1/2}t/\hbar G_n\}.$$

Sketch the variation of S with t, showing in particular that $S = n\pi$ when $e\sqrt{(I^2 - I_0^2)}t/\hbar G_n = n\pi/2$. Hence show that the time average of dS/dt is

$$\left\langle\frac{dS}{dt}\right\rangle = 2e(I^2 - I_0^2)^{1/2}/\hbar G_n$$

and derive eq. (5.35) for the d.c. voltage in the case of current feed.

5.5 Assuming the identity of eq. (5.38) (which is sometimes used as the defining equation for the Bessel coefficients $J_n(C)$), show that the supercurrent does take the form of eq. (5.39).

5.6 Confirm that the spacing in voltage of the current steps in Fig. 5.17 is $\hbar\omega/2e$, as predicted by eq. (5.40). How does the variation in amplitude of the various steps compare with the Bessel-function dependence of eq. (5.42)? (You can calibrate the microwave intensity by arranging that the first zero of the $n = 0$ step is at the first zero of $J_0(2ev/\hbar\omega)$. Note that the intensity is quoted in decibels of power, i.e. relative values of v^2).

5.7 Assume that the supercurrent in a microbridge is $I_0 f(S)$, where $f(S)$ is given by the Fourier sum of eq. (5.45). Show that eq. (5.39) is replaced by

$$I = I_0 \sum_{n,\,n'} (-1)^n A_{n'} J_n\left(\frac{2evn'}{\hbar\omega}\right) \sin\left\{\left(n'\frac{2eV}{\hbar} - n\omega\right)t - n\theta + n'S_0\right\}$$

and hence derive the microbridge synchronization condition of eq. (5.46).

References

AMBEGAOKAR, V. and BARATOFF, A., 1963, *Phys. Rev. Lett.* **10**, 468 and **11**, 104.
ANDERSON, P. W., 1964, *Lectures on the Many Body Problem*, Ravello Spring School 1963, Vol. 2, ed. E. R. Caianiello. (Academic Press, New York).

ANDERSON, P. W., 1966, *Quantum Fluids* (ed. D. F. Brewer), p. 146 (North Holland, Amsterdam).

ANDERSON, P. W., 1967, *Progress in Low Temperature Physics*, **5**, 1.

COON, D. D. and FISKE, M. D., 1965, *Phys. Rev.* **138**, A744.

DAYEM, A. H. and WIEGAND, J. J., 1967, *Phys. Rev.* **155**, 419.

DE BRUYN OUBOTER, R. and DE WAELE, A.TH.A.M., 1971, *Progress in Low Temperature Physics* **6**, 243.

DONNELLY, R. J. 1965, *Phys. Rev. Lett.* **14**, 939.

DWIGHT, H. B., 1961, *Tables of Integrals and Other Mathematical Data*, 4th ed. (Macmillan, Toronto).

FEYNMAN, R. P., 1965, *Lectures on Physics*, Vol. 3, Chapter 21, (Addison-Wesley, New York).

FINNEGAN, T. F., DENENSTEIN, A. and LANGENBERG, D. N., 1970, *Phys. Rev. Lett.* **24**, 738.

FISKE, M. D., 1964, *Rev. Mod. Phys.*, **36**, 221.

FORGACS, R. L. and WARNICK, A., 1967, *Rev. Sci. Instr.* **38**, 214.

GRIMES, C. C., RICHARDS, P. L. and SHAPIRO, S., 1966, *Phys. Rev. Lett.* **17**, 431

GRIMES, C. C. and SHAPIRO, S., 1968, *Phys. Rev.* **169**, 397.

HAMILTON, C. A. and SHAPIRO, S., 1970, *Phys. Rev.* **B2**, 4494.

HULIN, J. P., LAROCHE, C., LIBCHABER, A. and PERRIN, B., 1972, *Phys. Rev.* **A5**, 1830.

JAKLEVIC, R. C., LAMBE, J., MERCEREAU, J. E. and SILVER, A. H., 1965, *Phys. Rev.* **140**, A1628.

JOSEPHSON, B. D., 1962, *Phys. Lett.* **1**, 251.

JOSEPHSON, B. D., 1965, *Adv. Phys.* **14**, 419.

KHORANA, B. M., 1969, *Phys. Rev.* **185**, 299.

KHORANA, B. M. and DOUGLASS, D. H., JR., 1968, *Proc. 22th Int. Conf. on Low Temp. Phys.*, St. Andrews 1968, Vol. 1, 169 (University of St. Andrews, Scotland).

LANGENBERG, D. N., SCALAPINO, D. J., TAYLOR, B. N., and ECK, R. E., 1965, *Phys. Rev. Lett.* **15**, 294.

LANGENBERG, D. N., SCALAPINO, D. J. and TAYLOR, B. N., 1966, *Proc. I.E.E.E.* **54**, 560.

LEIDERER, P. and POBELL, F., 1973, *Phys. Rev.* **A7**, 1130.

MAMALADZE, YU. G. and CHEISHVILI, O. D., 1966, *Zh. Eksp. i Teor. Fiz.* **50**, 169 (*Soviet Physics JETP* **23**, 112, 1966).

MCCUMBER, D. E., 1968, *J. Appl. Phys.* **39**, 3113.

MUSINSKI, D. L. and DOUGLASS, D. H., 1972, *Phys. Rev. Lett.* **29**, 1541.

PARKER, W. H., TAYLOR, B. N. and LANGENBERG, D. N., 1967, *Phys. Rev. Lett.* **18**, 287.

RICHARDS, P. L., 1970, *Phys. Rev.* **A2**, 1532.

RICHARDS, P. L. and ANDERSON, P. W., 1965, *Phys. Rev. Lett.* **14**, 540.

SCHROEN, W., 1968, *J. Appl. Phys.* **39**, 2671.

SCOTT, W. C., 1970, *App. Phys. Lett.* **17**, 166.

SHAPIRO, S., 1963, *Phys. Rev. Lett.* **11**, 80.

SILVER, A. H. and ZIMMERMAN, J. E., 1967, *Phys. Rev.* **157**, 317.

SOLYMAR, L., 1972, *Superconductive Tunnelling and Applications* (Chapman and Hall, London).

STEWART, W. C., 1968, *App. Phys. Lett.* **12**, 277.

TAYLOR, B. N., PARKER, W. H. and LANGENBERG, D. N., 1969, *The Fundamental Constants and Quantum Electrodynamics*, (Academic Press, New York.)

WALDRAM, J. R., PIPPARD, A. B. and CLARKE, J., 1970, *Phil. Trans. Roy. Soc.* **A268**, 265.

WEBB, W. W., 1972, *I.E.E.E. Transactions on Magnetics*, **8**, 51.

YANSON, I. K., SVISTUNOV, V. M. and DMITRENKO, I. M., 1965, *Zh. Eksper. i Teor. Fiz.* **48**, 976 (*Soviet Physics JETP* **21**, 650).

ZIMMERMAN, J. E., 1970, *J. Appl. Phys.* **41**, 1589.

ZIMMERMAN, J. E., 1971a, *J. Appl. Phys.* **42**, 30.

ZIMMERMAN, J. E., 1971b, *J. Appl. Phys.* **42**, 4483.

ZIMMERMAN, J. E., THIENE, P. and HARDING, J. T., 1970, *J. Appl. Phys.* **41**, 1572.

Chapter 6

Ginzburg–Landau Theory

6.1 Introduction

One of the most fertile approaches to superconductivity has developed from a paper by Ginzburg and Landau (1950); we refer to the paper as GL. Before that time, Landau had developed a general theory for second-order phase transitions, based on the idea that a phase transition could be characterized by some kind of order parameter, and on a simple postulated form for the dependence of the free energy on the order parameter. In fact the postulated form does not always apply, but it is valid for superconductors. The crucial insight in GL was that for a superconductor the order parameter must be identified with the macroscopic wave function ψ; this means first that the order parameter is complex, and secondly that in general it varies in space. Once the free energy has been written down as a function of ψ and the vector potential $\mathbf{A}$, the equations for it to be a minimum give an equation of motion for ψ, and an equation for the supercurrent in terms of $\mathbf{A}$. The latter has the form of the London equation, so that the GL theory represents a generalization of the London theory to situations in which ψ is spatially varying.

The GL theory was developed before the microscopic theory. However, it was later shown, first by Gorkov (1959a), that in certain domains of temperature and magnetic field the GL equations are a rigorous consequence of the microscopic theory; the regions of validity of the GL equations are nowadays very well known. Since the GL equations are much simpler than the microscopic theory, they are generally used when they are known to hold. In addition, the GL equations give valuable insight into the general qualitative behaviour of superconductors, and of course they have sometimes been applied outside their strict range of validity. The theory is perhaps particularly useful in giving a clearer grasp of the relationship between the various lengths (penetration depth, coherence lengths) involved in superconductivity.

Naturally, the GL theory has also been applied to superfluidity in He II; the resulting equations are known as the Ginzburg–Pitaevskii (1958), or GP equations. However, there is a crucial difference between He II and superconductors which makes the theory much less useful for He II. The λ-anomaly means that the transition to superfluidity is by no means a perfect second-order phase transition, and in consequence the theory does not apply in a rigorous sense. Thus one is left with some valuable insight into superfluid behaviour in He II, but without any temperature range in which the theory is strictly valid.

6.2 The Landau theory of second-order phase transitions

The basic idea of the Landau theory is that a phase transition can be regarded as going from an ordered to a disordered phase. Consider, for simplicity, a model three-dimensional magnetic system consisting of an array of dipoles. If the interaction between the dipoles is ferromagnetic, that is, tends to align the dipoles, then the ground state at $T = 0$ has all the dipoles aligned. As T is increased, some dipoles go out of alignment because of thermal agitation. To be more precise, at a finite temperature T the state of the system is such as to minimize the Helmholtz free energy $F = U - T\Sigma$, where U is the internal energy. In the ground state, U is a minimum but Σ is zero. For $T > 0$, some spins go out of alignment in such a way as to increase both Σ and U and minimize F. We can define an order parameter ϕ as

$$\phi = \frac{n_+ - n_-}{n_+ + n_-} \tag{6.1}$$

where n_+ is the number of spins pointing upwards, and n_- the number pointing downwards. At $T = 0$, n_- is zero and $\phi = 1$. As T increases, ϕ decreases, becoming zero at the critical temperature T_c. Above T_c, in the paramagnetic phase, $n_+ = n_-$ and $\phi = 0$. The dependence of ϕ on T is sketched in Fig. 6.1. This type of behaviour is rather general in systems undergoing a phase change; for example in a superconductor the energy gap Δ, and in He II the superfluid density ρ_s, serve as order parameters. We recall that both Δ and ρ_s vary with temperature as in Fig. 6.1.

The Landau theory is concerned with the temperature region near T_c in which ϕ is small. The stable phase is the one in which the free energy F is a minimum. The first basic assumption of the theory is that in the region where ϕ is small F can be expanded as a power series in ϕ:

$$F = F_n + \lambda\phi + \alpha\phi^2 + \gamma\phi^3 + \tfrac{1}{2}\beta\phi^4. \tag{6.2}$$

There are systems in which such a simple expansion is not possible, but it does work for superconductivity. In eq. (6.2), all the coefficients, λ, α etc., are to be taken as functions of T, since the equilibrium values of F and ϕ are functions of T. The second basic assumption is that the coefficients can be expanded in powers of

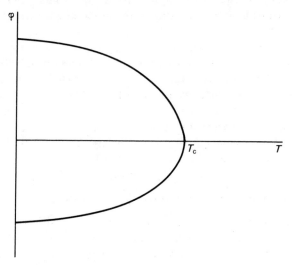

FIG. 6.1   Typical variation of an order parameter with temperature.

$T - T_c$. Again this simple assumption does not hold for all systems, but again it does hold for superconductivity. We can use this assumption to simplify eq. (6.2) considerably. First, the equilibrium phase corresponds to a minimum in F:

$$\frac{\partial F}{\partial \phi} = 0. \qquad (6.3)$$

In the normal phase, we must have a minimum at $\phi = 0$, which implies $\lambda = 0$ for $T > T_c$. Since we are assuming that λ can be expanded as a power series in $T - T_c$, this means we must have $\lambda = 0$ for all T. Furthermore, in most systems, including superconductors and He II, the term in ϕ^3 does not occur. We are thus left with

$$F = F_n + \alpha(T)\phi^2 + \tfrac{1}{2}\beta(T)\phi^4. \qquad (6.4)$$

where we now show the temperature dependences explicitly.

We now find the temperature dependence of α and β. Equation (6.3) for the value, ϕ_0, of ϕ at the minimum gives

$$\alpha\phi_0 + \beta\phi_0^3 = 0 \qquad (6.5)$$

with solutions $\phi_0 = 0$ and $\phi_0^2 = -\alpha/\beta$. Now we want $\phi_0 = 0$ to be the only solution for $T > T_c$, whereas for $T < T_c$ we must have a solution with $\phi_0 \neq 0$. We can achieve this if we take the temperature dependence so that $-\alpha/\beta$ is negative for $T > T_c$, and positive for $T < T_c$. In addition, we must have β positive at all temperatures, since if β were negative F would decrease indefinitely for large values of ϕ. We therefore want α to be positive for $T > T_c$, and negative for $T < T_c$. The simplest temperature dependences that give this are

$$\alpha(T) = A(T - T_c) \qquad (6.6)$$
$$\beta(T) = \beta(T_c) = \beta \qquad (6.7)$$

as sketched in Fig. 6.2, where A and β are positive constants.

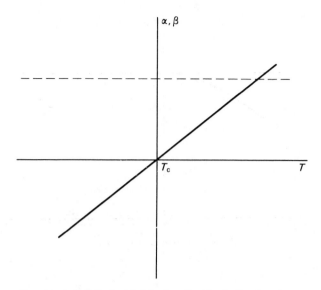

FIG. 6.2 α (full line) and β (broken line) in the Landau theory.

With α and β given by eqs. (6.6) and (6.7), the solution for ϕ_0, from eq. (6.5), is

$$\phi_0 = 0, \qquad\qquad\qquad\qquad T > T_c$$
$$\phi_0 = A^{1/2}(T_c - T)^{1/2}/\beta^{1/2}, \qquad T < T_c . \qquad (6.8)$$

In addition, $\phi_0 = 0$ remains a solution below T_c, but as we shall see shortly, it corresponds there to a maximum, not a minimum. Equation (6.8) gives a rapid, parabolic, increase in ϕ_0 as T decreases from T_c; this is as we sketched in Fig. 6.1. It is instructive to consider the free energy F as well as ϕ. First, the value of F at the minimum, F_{min}, is given by substituting ϕ_0 in eq. (6.4):

$$F_{min} = F_n, \qquad\qquad\qquad\qquad\qquad T > T_c$$
$$F_{min} = F_n - \tfrac{1}{2}\alpha^2/\beta = F_n - \tfrac{1}{4}A^2(T_c - T)^2/\beta, \qquad T < T_c . \qquad (6.9)$$

We sketch F_{min} as a function of T in Fig. 6.3. Note, in particular, that F_{min} decreases rather slowly from F_n as T decreases below T_c. The solution $\phi_0 = 0$ below T_c gives $F = F_n$; as we said, it gives a maximum, not a minimum. The fact that F_{min} changes only slowly, while ϕ_0 changes rapidly, with T, means that a thermal fluctuation which involves a large change in ϕ_0 need only require a small change of free energy. This is the reason for the sensitivity of second-order phase changes to fluctuation effects.

Besides the variation of F_{min} with T, it is instructive to look at the variation of F with ϕ implied by eq. (6.4). The dependence is sketched, for various temperatures, in Fig. 6.4. For T well above T_c, F is simply a parabolic-shaped curve, with a well-defined minimum. As T approaches T_c, and $\alpha(T)$ decreases, the minimum

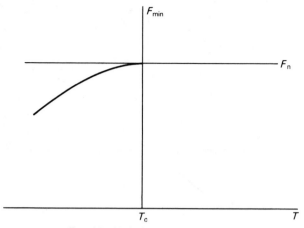

Fɪɢ. 6.3   Variation of F_{min} with T.

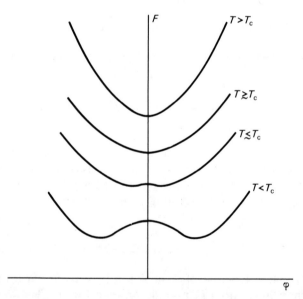

Fɪɢ. 6.4   Variation of F with ϕ at various temperatures (marked). The curves are displaced vertically from one another.

becomes shallower. When T passes through T_c, and $\alpha(T)$ becomes negative, the minimum at $\phi = 0$ becomes a local maximum, and minima develop at $\phi = \pm\phi_0$. As T decreases further, the minima move out in ϕ and slowly become deeper. F always increases for large enough ϕ, as we ensured by taking β to be positive.

This outline covers all we need of the general Landau theory. The key items are the dependence of F on ϕ, eq. (6.4) and Fig. 6.4, and the temperature dependences of ϕ_0 and F_{min}, Figs. 6.1 and 6.3.

6.3 Ginzburg–Landau equations

As we said in §6.1, the extension of the Landau theory to superconductors involves treating the wave function ψ as an order parameter. This introduces two complications: first, the order parameter becomes a function of position, in general; second, we must include explicitly the coupling of the supercurrent to the magnetic field and the magnetic-field energy. Both of these complications arise in the treatment of the mixed-state vortex lattice, for example.

Let us first make ψ a function of position; we can do this by treating eq. (6.4) as an expression for the free-energy density at the point $\mathbf{r}$, which we integrate over the volume of the specimen to get the total free energy. Furthermore, with ψ a function of position, we can expect a 'kinetic-energy' term in the energy proportional to $|\nabla\psi|^2$. The free-energy density is therefore

$$f(\mathbf{r}) = f_n + \alpha\,|\,\psi(\mathbf{r})\,|^2 + \frac{1}{2}\,\beta\,|\,\psi(\mathbf{r})\,|^4 + \frac{\hbar^2}{2m}\,|\,\nabla\psi(\mathbf{r})\,|^2. \tag{6.10}$$

We use modulus signs because ψ is complex. Following the usual convention in GL theory, we write the coefficient of the gradient term as $\hbar^2/2m$, where m is the mass of the electron. As de Gennes (1966) emphasizes, there is no physical content to this choice; in the end it simply determines the normalization of ψ. In fact, we did effectively make a different choice in Chapter 3, eq. (3.18), when we wrote London's equation using $2m$, the mass of a Cooper pair. We allowed for that by normalising ψ so that $|\psi|^2$ was $n_s/2$, the density of Cooper pairs. Here, therefore, we should interpret $|\psi|^2$ as n_s, the electron density, although in fact we shall not use this explicitly. We can take the coefficient of $|\nabla\psi|^2$ independent of T, like β, because we are dealing with a small temperature region near T_c, and the $|\nabla\psi|^2$ term must always be positive.

Even at this stage, we can see the principal physical consequence of adding the gradient term: it prevents ψ from changing too rapidly, since a high value of the gradient would give a large contribution to the free energy. On dimensional grounds, we can expect an appropriate ratio of coefficients to define a fundamental length: $\xi(T) = (\hbar^2/2m\,|\,\alpha\,|)^{1/2}$ is central to GL theory. It is clear that variations of ψ which are rapid within a distance $\xi(T)$ will not occur. We shall find that $\xi(T)$ is the coherence length for variations of the order parameter in the sense which we discussed in Chapter 4; for example, it is the core radius of a vortex line.

The inclusion of magnetic-field terms in the free energy requires a little care. The total induction $\mathbf{B}$ in the superconductor is the sum of the induction produced by the applied field $\mathbf{H}_0$, and that produced by the supercurrents $\mathbf{J}_e$:

$$\mathrm{curl}\,(\mathbf{B} - \mu_0\mathbf{H}_0) = \mu_0\mathbf{J}_e. \tag{6.11}$$

We extend the free energy of eq. (6.10) to include magnetic-field effects by making the usual replacement

$$\nabla \to \nabla \pm \frac{2ie}{\hbar}\,\mathbf{A}$$

with the $+$ sign if ∇ acts on ψ^*, and the $-$ sign if ∇ acts on ψ. We also add the magnetic-field energy, to get

$$f(\mathbf{r}) = f_n + \alpha |\psi(\mathbf{r})|^2 + \tfrac{1}{2}\beta |\psi(\mathbf{r})|^4$$

$$+ \frac{1}{2m} |(-i\hbar\nabla - 2e\mathbf{A})\psi|^2 + \mathbf{B}^2/2\mu_0 - \mu_0 H_0^2/2. \tag{6.12}$$

The integral of $f(\mathbf{r})$ over the volume of the specimen is the Helmholtz free energy $F = U - T\Sigma$. To maintain consistency with the thermodynamic equations of §1.2, we subtract $\tfrac{1}{2}\mu_0 H_0^2$, which is the magnetic energy of the coils generating the applied field $\mathbf{H}_0$. The internal energy U is the energy of the superconductor in the presence of the magnetic field, so that

$$dU = Td\Sigma + \mathbf{H}_0 \cdot d\mathbf{M}. \tag{6.13}$$

To find the stable state at temperature T and field $\mathbf{H}_0$, we must minimize the Gibbs free energy

$$G(T,\mathbf{H}_0) = U - T\Sigma - \mathbf{H}_0 \cdot \mathbf{M}. \tag{6.14}$$

Thus, finally, we have the result that we must minimize G:

$$G = \int g(\mathbf{r})d^3\mathbf{r} \tag{6.15}$$

with

$$g(\mathbf{r}) = f_n + \alpha |\psi(\mathbf{r})|^2 + \tfrac{1}{2}\beta |\psi(\mathbf{r})|^4$$

$$+ \frac{1}{2m} |(-i\hbar\nabla - 2e\mathbf{A})\psi|^2 + \mathbf{B}^2/2\mu_0 - \mathbf{H}_0 \cdot \mathbf{B} + \tfrac{1}{2}\mu_0 H_0^2. \tag{6.16}$$

In some accounts of GL theory, it is stated that the free energy $F = \int f(\mathbf{r})d^3r$ should be minimized; the Helmholtz energy density $f(\mathbf{r})$ differs from the Gibbs density $g(\mathbf{r})$ by $\mathbf{H}_0 \cdot \mathbf{M}$. We shall see that one gets the same equations for $\psi(\mathbf{r})$ and for the supercurrent whether one minimizes F or G. However, in discussing the parallel critical fields of thin films (§6.6), for example, it is important to use the correct function G.

The energy G is an integral involving two functions $\psi(\mathbf{r})$ and $\mathbf{A}(\mathbf{r})$ (recall that $\mathbf{B} = \text{curl } \mathbf{A}$). This contrasts with the ordinary Landau theory, where G is a function of the variable ϕ. In that case, the equation for a minimum was simply $\partial G/\partial\phi = 0$. Now, since G depends on functions ψ and $\mathbf{A}$, we must use the Euler–Lagrange equations of the calculus of variations. Since ψ is complex, we can minimize with respect to either ψ or ψ^*; the ψ^* equation is

$$\frac{\partial g}{\partial \psi^*} - \sum_j \frac{\partial}{\partial x_j} \frac{\partial g}{\partial(\nabla_j \psi^*)} = 0, \tag{6.17}$$

where $\nabla_j \psi^*$ is the component of the gradient in the direction j. This equation is sometimes written formally as $\delta G/\delta\psi^* = 0$, where $\delta G/\delta\psi^*$ stands for the left-hand side of eq. (6.17). After some manipulation, and with the restriction that we use the gauge div $\mathbf{A} = 0$, eq. (6.17) becomes

$$\frac{1}{2m}(-i\hbar\nabla - 2e\mathbf{A})^2\psi + \alpha\psi + \beta|\psi|^2\psi = 0. \tag{6.18}$$

This is the first GL equation. If we had minimized with respect to ψ rather than ψ^* we would simply have found the complex-conjugate equation.

We now have a differential equation which will give us the variation of ψ within any specimen once we know the vector potential $\mathbf{A}$. As always, we must supplement the differential equation with appropriate boundary conditions for ψ at the surface of the specimen. Ginzburg and Landau argued as follows. The complete expression for the variation δG of G when ψ^* varies by $\delta\psi^*$ comprises first the volume integral of the left-hand side of eq. (6.17), multiplied by $\delta\psi^*$, and secondly the surface integral

$$I_s = \int \delta\psi^* \frac{\partial g}{\partial(\nabla\psi^*)} \cdot \mathbf{n}\,dS \tag{6.19}$$

where $\mathbf{n}$ is the normal to the surface. In the usual formulation of the calculus of variations, this term is neglected because the variation is carried out with the subsidiary condition $\psi^* = 0$, and therefore $\delta\psi^* = 0$, at the surface. For superconductors, however, we cannot simply impose the condition $\psi^* = 0$ at a surface. Indeed, such a condition would be wrong: for example, it would mean that the critical temperature of a thin film oscillated as a function of its thickness, since a standing-wave condition would be involved. In fact the critical temperature does not oscillate, and is generally independent of film thickness. Ginzburg and Landau therefore took the boundary condition

$$\mathbf{n} \cdot \frac{\partial g}{\partial(\nabla\psi^*)} = 0, \tag{6.20}$$

which of course also makes I_s vanish. Written out explicitly, the boundary condition reads

$$\mathbf{n} \cdot (-i\hbar\nabla - 2e\mathbf{A})\psi = 0. \tag{6.21}$$

The derivation from the microscopic theory shows that this condition is correct for the boundary between a superconductor and an insulator. If a normal metal adjoins the superconductor, the wave function penetrates some distance into the normal metal, and because of this proximity effect the boundary condition at a superconducting normal interface differs from eq. (6.21). We note finally that eq. (6.21) does give a critical temperature independent of thickness for a thin film. In the absence of a magnetic field, the condition is simply that the slope of ψ is zero at the surface, which means that ψ is constant across the film at the critical temperature. Consequently, the critical temperature is unaffected by the thickness.

We must now minimize the free energy G with respect to variations of the vector potential $\mathbf{A}$. The appropriate Euler–Lagrange equation is

$$\frac{\partial g}{\partial A_i} - \sum_j \frac{\partial}{\partial x_j}\frac{\partial g}{\partial(\partial A_i/\partial x_j)} = 0, \tag{6.22}$$

where A_i is the component of $\mathbf{A}$ in the ith direction. The second term in this equation gives $(1/\mu_0)$ curl curl $\mathbf{A}$ or $(1/\mu_0)$ curl $\mathbf{B}$, which from Maxwell's equations is $\mathbf{J}_e$. With this replacement, eq. (6.22) becomes

$$\mathbf{J}_e = -\frac{ieh}{m}(\psi^* \nabla\psi - \psi \nabla\psi^*) - \frac{4e^2}{m} \psi^*\psi \mathbf{A}, \tag{6.23}$$

that is, the standard expression for a quantum-mechanical current. The fact that we get the standard expression is not surprising: $\mathbf{A}$ enters the free energy only in the gradient term and in the field-energy term $B^2/2\mu_0$, and that is exactly where we should find it in a quantum-mechanical Hamiltonian. Except that we are now using the mass m rather than $2m$, which means only a change in the normalization of ψ, eq. (6.23) is therefore identical to our earlier eq. (3.18), which we wrote down on plausibility grounds from the assumption that a superconductor is characterized by a macroscopic wave function. Just like eq. (3.18), eq. (6.23) is a local expression, of the London type; $\mathbf{J}_e(\mathbf{r})$ is given by the values of $\nabla\psi$ and $\mathbf{A}$ at the point $\mathbf{r}$. The fact that we have a local theory implies a severe restriction on the temperature range in which the theory is valid if we are dealing with a pure superconductor. For an alloy, on the other hand, the temperature range need not be restricted by the local nature of the theory.

Note that, as we stated, we should have got the same expression for the current if we had used the Helmholtz density, eq. (6.12), rather than the Gibbs density, eq. (6.16). The two differ by the term $\mathbf{H}_0 \cdot \mathbf{B} = \mathbf{H}_0 \cdot$ curl $\mathbf{A}$; if we put this into the second term of eq. (6.22) we get curl $\mathbf{H}_0$, which is zero inside the superconductor.

We have now developed the full framework of the GL theory: the Gibbs free energy, eqs. (6.15) and (6.16); eq. (6.18) for ψ, with the boundary condition of eq. (6.21); and finally eq. (6.23) for the current. The temperature dependence of the parameters α and β remains that of the Landau theory:

$$\alpha = -A(T_c - T) \tag{6.24}$$

$$\beta = \text{constant.} \tag{6.25}$$

Our first task is to derive expressions for the penetration depth, and for the critical field H_{cb} of a bulk type I superconductor. If the dimensions of the specimen are much greater than λ, then we have $\mathbf{B} = 0$ inside the specimen. We then have ψ constant; if ψ varied the gradient term would mean that the free energy increased. The constant value of ψ is given from eq. (6.18):

$$|\psi|^2 = -\alpha/\beta = |\alpha|/\beta \tag{6.26}$$

(recall that α is negative), which of course is the same as in the ordinary Landau theory, eq. (6.8). Since $\nabla\psi = 0$, the current, eq. (6.23), is simply given by the London equation

$$\mathbf{J}_e = -\frac{4e^2}{m}\frac{|\alpha|}{\beta} \mathbf{A}. \tag{6.27}$$

As we saw in § 3.3, this gives for the penetration depth

$$\lambda = \left(\frac{m\beta}{4e^2\mu_0 |\alpha|}\right)^{1/2}. \tag{6.28}$$

Equations (6.24) and (6.25) lead to an explicit form for the temperature dependence of λ:

$$\lambda \propto (1-t)^{1/2}. \tag{6.29}$$

We saw earlier (Fig. 3.6) that the Gorter–Casimir temperature dependence $\lambda \propto (1-t^4)^{-1/2}$ fits experimental data at all temperatures. Near to T_c, which is where the Landau theory holds, the two forms of temperature dependence are in agreement. In fact, $(1-t^4)^{-1/2} = (1+t^2)^{-1/2}(1+t)^{-1/2}(1-t)^{-1/2}$, and for t near 1 the first two terms are slowly varying, so that the dependence on t is dominated by the singularity given by the last term.

H_{cb} is the field at which the Gibbs free energies G_s and G_n in the superconducting and normal phase are equal. $\mathbf{B}$ is zero in the superconducting phase, and ψ is zero in the normal phase. With $|\psi^2| = |\alpha|/\beta$, eq. (6.15) therefore gives for the energies at applied field $\mathbf{H}_0$

$$G_s = V(f_n - |\alpha|^2/2\beta + \tfrac{1}{2}\mu_0 H_0^2) \tag{6.30}$$

$$G_n = Vf_n, \tag{6.31}$$

where V is the volume of the specimen. The critical field is given by $G_s = G_n$:

$$H_{cb}^2 = |\alpha|^2/\mu_0\beta \tag{6.32}$$

which is of course the usual result that at the critical field the flux-exclusion energy is equal to the condensation energy. Equation (6.32) gives the temperature dependence $H_{cb} \propto 1-t$, which again for t near 1 is in agreement with the Gorter–Casimir form $1-t^2$, and therefore in agreement with the experimental results. As Ginzburg and Landau remarked, this is strong support for the postulated temperature dependences of α and β, eqs. (6.24) and (6.25).

The phase change at H_{cb} is first order, that is, there is a latent heat associated with the phase change. Within the Landau theory, the nature of a given phase change is connected with the behaviour of the order parameter ψ at the phase boundary. In the present case, ψ retains the constant value $(|\alpha|/\beta)^{1/2}$ up to H_{cb}, and then drops discontinuously to zero. This behaviour is sketched in Fig. 6.5. The

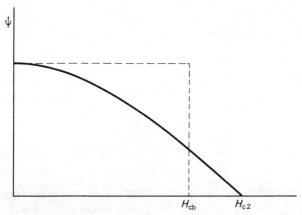

FIG. 6.5   Variation of ψ with H in bulk type I superconductor (broken line), and expected variation in type II superconductor (full line).

discontinuous change of ψ means that there is a finite difference in the degree of order between the two phases. This implies an entropy difference $\Delta\Sigma$ and an associated latent heat $T\,\Delta\Sigma$. Since the superconducting phase is more ordered, it has the lower entropy, so if the experiment is done isothermally heat is absorbed from the surroundings in the passage from the superconducting to the normal phase. An expression for the latent heat can of course be obtained from eq. (6.30); we leave the explicit evaluation for Problem 6.4.

6.4   Second-order critical fields

We have already seen, in § 1.2 and Chapter 4, that whereas a type I superconductor undergoes a first-order transition into the normal state at the critical field H_{cb}, a bulk type II superconductor undergoes a second-order transition at a critical field H_{c2}. At H_{cb}, the magnetization drops discontinuously to zero, whereas at H_{c2} the approach to zero is continuous. This means that at H_{c2} the supercurrent density J_e approaches zero continuously, and consequently ψ approaches zero continuously (eq. 6.23). We have seen that in the type I superconductor, on the other hand, ψ has a constant value up to H_{cb}, and then drops abruptly to zero. The variation of ψ with magnetic field in the two cases is sketched in Fig. 6.5. The graph for the mixed state is meant in the following sense: ψ is a function of position in the mixed state, which contains circulating supercurrents, and we draw the variation with field of some average value of ψ. Any average value of ψ tends to zero at H_{c2}.

We can considerably simplify the GL equation for ψ, eq. (6.18), for the purpose of calculating a second-order critical field like H_{c2}. Just below H_{c2}, ψ is small everywhere, and consequently the non-linear term $\beta\,|\,\psi\,|^2\psi$ is small compared with the other terms. Furthermore, the circulating supercurrents, which are proportional to $|\,\psi\,|^2$, are small, and so the magnetic field $\mathbf{H}$ which comes into eq. (6.18) is close to the uniform applied field $\mathbf{H}_0$. Just at H_{c2}, we can ignore $\beta\,|\,\psi\,|^2\psi$ and the difference $\mathbf{H}-\mathbf{H}_0$. We therefore have the following prescription for calculating a second-order critical field: linearize the equation for ψ by dropping the term $\beta\,|\,\psi\,|^2\psi$, and use the vector potential $\mathbf{A}$ appropriate to a uniform applied field $\mathbf{H}_0$. In general, we shall also have to use the boundary condition of eq. (6.21) for any free surfaces which are present.

Let us see how this prescription works for calculating the bulk critical field H_{c2}. This is the easiest case, because there are no surfaces involved. We take the z-axis along the uniform applied field $\mathbf{H}_0$:

$$\mathbf{H}_0 = (0, 0, H_0). \tag{6.33}$$

There is considerable freedom in the choice of gauge for $\mathbf{A}$; we take

$$\mathbf{A} = (0, \mu_0 H_0 x, 0). \tag{6.34}$$

An alternative would be $(-\mu_0 H_0 y, 0, 0)$, and indeed any linear combination of this with eq. (6.34) would serve too. All these choices satisfy div $\mathbf{A} = 0$ and give $\mu_0\mathbf{H}_0 = \text{curl}\,\mathbf{A}$; the calculated value of H_{c2} is naturally independent of the choice of gauge.

With eq. (6.34), the linearized form of eq. (6.18) becomes

$$-\frac{\hbar^2}{2m}\frac{\partial^2\psi}{\partial x^2}+\frac{1}{2m}\left(-i\hbar\frac{\partial}{\partial y}-2e\mu_0 H_0 x\right)^2\psi-\frac{\hbar^2}{2m}\frac{\partial^2\psi}{\partial z^2}=|\alpha|\psi. \tag{6.35}$$

It can be seen that this is the Schrödinger equation for a particle of mass m and charge $2e$ moving in a uniform magnetic field directed along the z-axis. $|\alpha|$ plays the part of the eigenvalue. We can describe the possible trajectories classically as spiral paths in which the particle moves along the field direction with a constant momentum $\hbar k_z$, and rotates about the field direction with a constant angular velocity. The quantum-mechanical energy eigenvalues are

$$|\alpha| = (n+\tfrac{1}{2})\hbar\omega_c + \hbar^2 k_z^2/2m, \tag{6.36}$$

where ω_c is the cyclotron resonance frequency:

$$\omega_c = 2e\mu_0 H_0/m. \tag{6.37}$$

The derivation of these results proceeds by reducing eq. (6.35) to the equation for the quantum-mechanical harmonic oscillator. We observe first that the coordinates y and z enter only via the derivatives $\partial/\partial y, \partial/\partial z$, which implies that the corresponding parts of the eigenfunction are plane waves:

$$\psi(\mathbf{r}) = f(x)\exp(ik_y y)\exp(ik_z z), \tag{6.38}$$

where $f(x)$ is yet to be determined. With this substitution, eq. (6.35) becomes

$$-\frac{\hbar^2}{2m}\frac{d^2f}{dx^2}+\frac{1}{2m}(\hbar k_y-2e\mu_0 H_0 x)^2 f=\left(|\alpha|-\frac{\hbar^2 k_z^2}{2m}\right)f. \tag{6.39}$$

We note immediately that k_y does not enter into the eigenvalue; thus any given eigenvalue is highly degenerate because the corresponding eigenfunction can contain any value of k_y. The different values of k_y correspond to differe... centres x_0 for the eigenfunction $f(x)$. We put

$$\hbar k_y = 2e\mu_0 H_0 x_0 \tag{6.40}$$

and then eq. (6.39) becomes

$$-\frac{\hbar^2}{2m}\frac{d^2f}{dx^2}+\frac{2e^2\mu_0^2 H_0^2}{m}(x-x_0)^2 f=(|\alpha|-\hbar^2 k_z^2/2m)f. \tag{6.41}$$

This asymmetry between x and y is simply a consequence of our initial choice of an asymmetric gauge, eq. (6.34). If we had chosen the symmetric gauge $(-\tfrac{1}{2}\mu_0 H_0 y, \tfrac{1}{2}\mu_0 H_0 x, 0)$ the treatment would have been symmetric.

Equation (6.41) is the Schrödinger equation for a particle moving in a harmonic potential well centred at x_0, and the eigenvalues are indeed those given by eqs. (6.36) and (6.37). We must read the equations in a slightly unusual way, however. Usually one knows the magnetic field H_0 and wishes to find the eigenvalue $|\alpha|$; but we know $|\alpha|$ and we wish to know H_0. We therefore turn the equations round:

$$H_0 = \frac{m}{(2n+1)e\hbar\mu_0}\left(|\alpha|-\frac{\hbar^2 k_z^2}{2m}\right). \tag{6.42}$$

This equation is far too general, since we are concerned only with the largest H_0-value; at any lower field the non-linear terms will have come into play, and our initial premise that we could neglect them will not hold. The highest value of H_0 is H_{c2}, and it is obviously given by $n = 0$ and $k_z = 0$:

$$H_{c2} = m|\alpha|/e\hbar\mu_0. \tag{6.43}$$

The corresponding eigenfunction is the ground-state wave function of the harmonic oscillator:

$$f(x) = \exp\{-(x - x_0)^2/2\xi^2(T)\}, \tag{6.44}$$

where

$$\xi^2(T) = \hbar^2/2m|\alpha|. \tag{6.45}$$

The potential well which comes into eq. (6.41) and the wave function are shown in Fig. 6.6. As we anticipated when we introduced the gradient term in the free

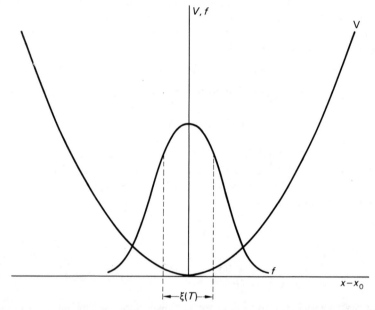

FIG. 6.6   Harmonic oscillator potential $V(x - x_0)$ and lowest eigenfunction $f(x - x_0)$.

energy, $\xi(T)$ is the characteristic distance for variations in the order parameter. We can rewrite eq. (6.43) as

$$H_{c2} = \phi_0/2\pi\mu_0\xi^2(T), \tag{6.46}$$

which we have already quoted in §4.3.

As we saw in §4.3, $\xi(T)$ is related to $\lambda(T)$, and H_{c2} to H_{cb}, by the parameter κ. The relevant equations are (6.28) for λ, (6.44) for ξ, (6.32) for H_{cb} and (6.43) for H_{c2}. Comparing these, we find

$$H_{c2} = \kappa\sqrt{2}H_{cb}, \tag{6.47}$$

$$\lambda(T) = \kappa\xi(T), \tag{6.48}$$

with

$$\kappa^2 = m^2\beta/2\mu_0 e^2\hbar^2. \tag{6.49}$$

Within the framework of the GL theory, κ is independent of temperature, which is to say that the two fields H_{c2} and H_{cb}, and the two lengths $\xi(T)$ and $\lambda(T)$, have similar temperature dependences.

The interpretation of H_{c2} for a type II superconductor, with $\kappa > 1/\sqrt{2}$, is quite clear: it is the boundary of the mixed state. In a type I superconductor, with $\kappa < 1/\sqrt{2}$ and $H_{c2} < H_{cb}$, H_{c2} has the significance of a critical supercooling field, as we shall now see. The phase transition at H_{cb} is first order; at H_{cb} the Gibbs energies are equal, and there is a finite difference in the order parameter ψ between the two phases. As shown schematically in Fig. 6.7, both phases represent a local minimum in G. As the field is reduced below H_{cb}, G_s decreases, while G_n is

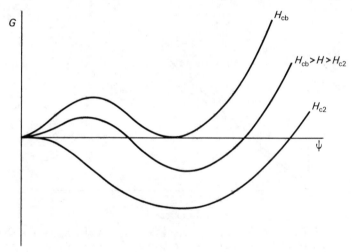

FIG. 6.7   Schematic variation of G with ψ for various fields in a type I superconductor. The normal phase corresponds to a minimum in G at $\psi = 0$, and the superconducting phase to a minimum in G at a non-zero value of ψ. At H_{cb}, the values of G at both minima are equal, $G_n = G_s$. For a range of fields below H_{cb}, $G_n > G_s$, but the normal phase still corresponds to a local minimum of G.

unaltered (eqs. 6.30 and 6.31). For some range of fields, the normal phase still represents a local minimum, that is, it is a metastable phase. It is therefore possible for the normal phase to persist for fields below H_{cb}; the phenomenon is called supercooling, because of the obvious analogy with a liquid–vapour system. Our calculation of H_{c2}, starting from the linearized equation for ψ, shows that H_{c2} is the lowest field for which the equation $\delta G/\delta\psi = 0$ has a solution for a small, but non-zero, value of ψ. In terms of Fig. 6.7, therefore, H_{c2} corresponds to a flat dependence of G on ψ near $\psi = 0$. This means that H_{c2} is the field at which the local minimum at $\psi = 0$ turns into a local maximum; H_{c2} is therefore the boundary of stability of the normal phase, which is what is meant by a critical supercooling field.

Experiments on supercooling of bulk type I superconductors were first carried out by Faber (1952, 1955, 1957). He found that well below T_c the degree of supercooling possible varied from specimen to specimen and from place to place in a given specimen. Faber attributed this to the fact that the superconducting phase appears at a specific nucleation site, and then grows throughout the specimen. Near to T_c, however, the supercooling becomes independent of the particular specimen; as Faber remarked, this occurs because near to T_c the coherence length is so long that the volume which first turns superconducting, of order $\xi^3(T)$, is much larger than any defect in the specimen. Consequently the defects do not matter, and the nucleation becomes a property of the specimen as a whole. An example of the ideal supercooling near to T_c is shown in Fig. 6.8. We shall see later

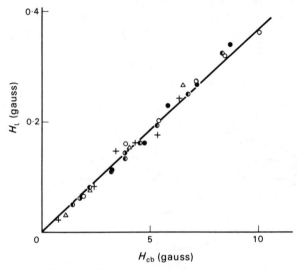

FIG. 6.8   Ideal supercooling for Al. The results are plotted as the supercooling field H_1 versus H_{cb}, with temperature as a parameter. The data are taken from three sections of each of two specimens. (After Faber, 1957.)

that in fact the supercooling field H_1 in Faber's work is probably the surface field H_{c3}, rather than H_{c2}. Because of the difficulty with nucleation sites, supercooling studies have never been carried out far below T_c in bulk specimens, and more recent experimental work, e.g. Feder and McLachlan (1969), has been on small specimens, particularly spheres.

One can also define a superheating field as the limit of metastability of the superconducting phase. Experiments on superheating are made difficult, however, by the fact that the flux is excluded from the superconducting phase, so that the field at some parts of the specimen surface is larger than the applied field (Fig. 1.16). Doll and Graf (1967) overcame this difficulty by applying the field only to the middle part of a cylindrical specimen. They prepared their specimens with great care; some results for the superheating field H_{sh} and H_{cb} are shown in Fig. 6.9.

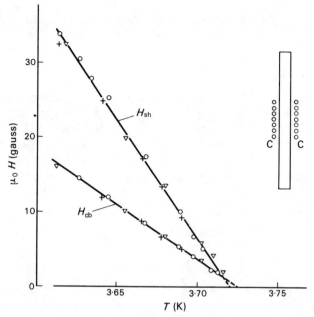

FIG. 6.9 Superheating field H_{sh} in three different Sn specimens. The field was applied only over the middle part of the specimen, from coil C. This technique ensures that no supercooling occurs. The critical fields were derived from ascending and descending field magnetization curves. (After Doll and Graf, 1967.)

Having dealt with the bulk second-order critical field H_{c2}, we now turn to the calculation of critical fields in the presence of surfaces. The simplest case is when we have an applied field parallel to the surface of a bulk specimen, and the next simplest is a thin film in a parallel applied field.

Consider first a specimen with a single plane surface in a parallel field. We take the z-axis in the direction of the field, and the x-axis as the normal to the surface, Fig. 6.10. We take $\mathbf{A} = (0, \mu_0 H_0 x, 0)$ as in eq. (6.34), so that we still have to solve eq. (6.35). Now, however, we have to take account of the boundary condition, eq. (6.21), which reduces in the present case to

$$\frac{\partial \psi}{\partial x} = 0 \qquad \text{at} \qquad x = 0. \tag{6.50}$$

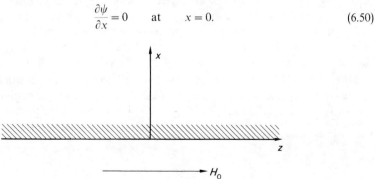

FIG. 6.10 Coordinate system for calculation of surface critical field.

The treatment of eq. (6.35) runs in parallel with what we did for the bulk case; we take ψ in the form of eq. (6.38), and arrive at

$$-\frac{\hbar^2}{2m}\frac{d^2 f}{dx^2} + \frac{2e^2\mu_0^2 H_0^2}{m}(x - x_0)^2 f = |\alpha| f. \tag{6.51}$$

This is the same as eq. (6.41), except that we have omitted the term in k_z^2, since that decreases H_0 for a given $|\alpha|$. We still have the Schrödinger equation for a particle in a harmonic well centred at x_0; now, however, the boundary condition, eq. (6.50), means that the eigenvalue depends crucially on the value of x_0. As was pointed out by Saint-James and de Gennes (1963), who first formulated and solved this problem, one can take account of the boundary condition by an image method.

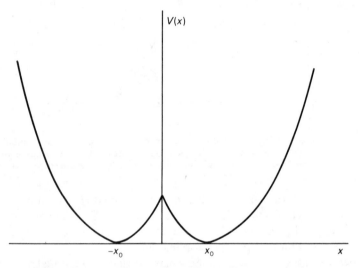

FIG. 6.11   Potential well for solving surface problem by image method.

Consider a particle moving in the potential $V(x)$ of Fig. 6.11, which is our original harmonic well together with its reflection in the surface. The ground-state wave function $\psi_0(x)$ in $V(x)$ must be symmetric about $x = 0$; this means that it has zero slope at $x = 0$, as required in eq. (6.50), and for $x > 0$ it satisfies eq. (6.51). Thus $\psi_0(x)$ for $x > 0$ is the solution to our problem, and the corresponding eigenvalue E_0 gives the critical field. We can compare E_0 for various x_0 with the eigenvalue E in the single harmonic well. For $x_0 \to \infty$, $E_0 \to E$, and for $x_0 = 0$, $E_0 = E$ again. For intermediate values of x_0, E_0 is less than E because $V(x)$ is smaller than the harmonic potential in some regions. The detailed solution can be given in terms of Weber functions, which satisfy eq. (6.51). It is found that the lowest eigenvalue is given by

$$x_0 = 0.59 \xi(T) \tag{6.52}$$

$$|\alpha| = 0.59(\tfrac{1}{2}\hbar\omega_c) \tag{6.53}$$

where ω_c is the cyclotron resonance frequency, and $\frac{1}{2}\hbar\omega_c$ is the eigenvalue in the harmonic well. We can turn eq. (6.53) round to get the critical field:

$$H_{c3} = 1\cdot69\kappa\sqrt{2}H_{cb} = 1\cdot69H_{c2}. \qquad (6.54)$$

Thus at H_{c3}, superconductivity appears in the form of a surface sheath with a thickness about $\xi(T)$ on surfaces parallel to the applied field.

Since its original prediction, the existence of the surface sheath has been confirmed in a great variety of experiments. The most direct method is to measure both the magnetic and resistive transitions of a bulk specimen. Between H_{c2} and H_{c3} a supercurrent can flow in the surface sheath, so that the resistive transition for a small measuring current occurs at H_{c3}. The magnetization, on the other hand, is governed by the behaviour of the bulk of the specimen, so the magnetic transition occurs at H_{c2}. The temperature dependence of critical fields measured by this method is shown in Fig. 6.12. An alternative is to measure the critical

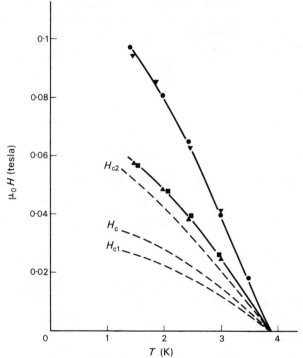

FIG. 6.12   Critical fields for In–6% Pb alloy. H_{c1}, H_c and H_{c2} from magnetization measurements. H_{c3} for annealed and copper plated samples: ● before copper plating; ▼ after removing copper; ▲ copper plated (5×10^{-5} m thick); ■ copper plated (2×10^{-5} m thick). (After Gygax and Kropschot, 1964.)

perpendicular and parallel fields $H_{c\perp}$ and $H_{c\parallel}$ of a thick alloy (type II) film. If the thickness d is much greater than $2\xi(T)$, the surfaces of the film are independent, and $H_{c\parallel}$ is simply H_{c3}. The perpendicular critical field is H_{c2}. We therefore have

$$H_{c\parallel} = 1\cdot69H_{c\perp} \qquad \text{for} \qquad d \gg 2\xi(T). \qquad (6.55)$$

Figure 6.14 shows critical fields of various In alloy films, taken from tunnelling measurements, and it can be seen that eq. (6.55) holds in the thick limit.

In a type I superconductor which has $H_{c3} < H_{cb}$, H_{c3} may be expected to appear as the critical supercooling field, so that in Faber's experiments, for example, the supercooling field was probably H_{c3} rather than H_{c2}. It is also possible, of course, to have intermediate-type superconductors with $H_{c2} < H_{cb} < H_{c3}$, in which the surface sheath appears in some field interval above H_{cb}.

Surface sheath effects can be suppressed if the superconducting surface is coated with a normal metal. An example is shown in Fig. 6.12, where copper-plated specimens have a resistive critical field very close to H_{c2}. We mentioned earlier that the boundary condition of zero gradient does not hold for the interface with a normal metal; the consequence of using the correct boundary condition is indeed that the critical field is reduced below H_{c3}. In fact the exact calculation by Hurault (1966) shows that the critical field is exactly H_{c2} if the conductivity of the normal metal is greater than the normal-state conductivity of the superconductor. It is standard practice nowadays to suppress unwanted surface sheath effects by coating with a layer of normal metal.

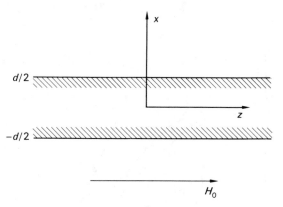

FIG. 6.13   Coordinate system for a film of thickness d in a parallel field.

We now turn to the second of our problems concerning boundaries, namely the critical field of a film in a parallel field. We take the coordinate system of Fig. 6.13, with the film boundaries at $x = \pm \frac{1}{2}d$. We have to solve eq. (6.51), now with the boundary conditions

$$\frac{df}{dx} = 0 \quad \text{at} \quad x = \pm \frac{1}{2}d. \tag{6.56}$$

The character of the solution clearly depends on the ratio $d/\xi(T)$. For $d \gg \xi(T)$, as we have already pointed out, the two surfaces are effectively independent, and we get sheath nucleation on both surfaces at the critical field H_{c3}. For $d \sim \xi(T)$, the situation is complex, and the problem has to be solved in terms of Weber functions. For $d \ll \xi(T)$, however, the problem simplifies. In this thin limit, f can hardly vary across the film, and we can integrate eq. (6.51) across the thickness of the film. The first term gives

$$-\frac{\hbar^2}{2m}\left\{\left(\frac{df}{dx}\right)_{d/2}-\left(\frac{df}{dx}\right)_{-d/2}\right\},$$

which is zero because of the boundary conditions. In integrating the other two terms, we can take f as a constant, f_0, to get

$$\frac{2e^2\mu_0^2 H_0^2}{m}\left(\frac{d^3}{12}+x_0^2 d\right)f_0 = |\alpha|f_0 d. \tag{6.57}$$

We are interested in the smallest eigenvalue $|\alpha|$, or equivalently in the largest critical field H_0. This obviously requires taking $x_0 = 0$ in eq. (6.57), which means, not surprisingly, that the potential well for the problem is symmetric. With this choice, eq. (6.57) gives for the critical field

$$H_{c\parallel}^2 = \frac{6m|\alpha|}{\mu_0^2 e^2 d^2} \qquad \text{for} \qquad d \ll \xi(T). \tag{6.58}$$

It is convenient to refer the parallel critical field to the perpendicular critical field, which is H_{c2}, given by eq. (6.43). We define the reduced variables

$$\varepsilon = \mu_0 H_{c\perp} d^2 \pi/2\phi_0 = d^2/4\xi^2(T) \tag{6.59}$$

$$h = \mu_0 H_{c\parallel} d^2 \pi/2\phi_0. \tag{6.60}$$

The two limits which we have found are

$$h = 1{\cdot}69\varepsilon \qquad \varepsilon \gg 1 \tag{6.61}$$

$$h^2 = 3\varepsilon \qquad \varepsilon \ll 1. \tag{6.62}$$

The advantage of this representation is that the experimental points fall on the same universal curve of h against ε for all thicknesses and temperatures. As T increases for a given film, ε decreases because $\xi(T)$ increases. Some experimental results for critical fields of In alloy films are shown in Fig. 6.14. The theoretical

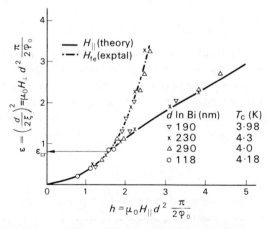

FIG. 6.14 Parallel and perpendicular critical fields of In alloy films at various temperatures, plotted in the reduced variables of eqs. (6.59) and (6.60). The theoretical curve (full line) is from the calculation of Saint-James and de Gennes (1963), with the asymptotic forms $h^2 = 3\varepsilon$, $h = 1{\cdot}69\varepsilon$ for small and large ε respectively. Also shown is the flux entry field, H_{fe}, which diverges from the critical field at $\varepsilon_{cr} = 0{\cdot}82$.
(After Guyon et al., 1967.)

curve for h versus ε is taken from the full calculation of Saint-James and de Gennes (1963), with the limiting values given by eqs. (6.61) and (6.62).

There is an important qualitative difference between the character of the solutions in the thin and thick limits; the difference is governed by the behaviour of the point x_0, the centre of the harmonic well in our Schrödinger equation. We have seen from eq. (6.58) that in the thin limit, $d \ll 2\xi(T)$, we have $x_0 = 0$. On the other hand, in the thick limit, the well is centred at a distance $0.59\xi(T)$ from either boundary, so that we have $x_0 = \pm(d/2 - 0.59\xi(T))$. The variation of x_0 with thickness is sketched in Fig. 6.15. The critical point at which x_0 deviates from zero is found from the exact calculation to be $\varepsilon_{cr} = 0.816$, $h_{cr} = 1.63$ in terms of the reduced variables; the critical field curve changes curvature at this point, as can be seen in Fig. 6.14. When x_0 is non-zero, there are two possible wave functions, one for $+x_0$, one for $-x_0$, corresponding to the lowest eigenvalue. These wave functions, which of course are asymmetric about $x = 0$, are sketched in Fig. 6.16. For $d \gg 2\xi(T)$ the two wave functions are simply the surface sheath functions on each surface.

The interesting result concerns the character of the wave function in a field just below the critical field. It is important to put in explicitly the dependence on the y-coordinate, which is in the plane of the film at right angles to the field; we recall that there is a phase factor $\exp(iky)$. Any linear combination of the two degenerate wave functions centred at x_0 and $-x_0$ is a solution of the linear GL equation. However, the value of the free energy, including the non-linear terms, depends on the relative weights of the two degenerate wave functions, and the correct linear combination is the one which minimizes the free energy. We shall not go into the details of the calculation of the free energy, which is similar to that given in the next section for the Abrikosov lattice. The result of the calculation is that the functions centred at x_0 and $-x_0$ have equal weight, so that the wave function may be written:

$$\psi(x,y) = A\{\exp(iky)f(x) + \exp(-iky)f(-x)\}, \tag{6.63}$$

with k and x_0 related by eq. (6.40). The amplitude A would eventually have to be chosen to minimize the free energy, including the non-linear terms, but we shall not go into that. The relative phase of the two terms in eq. (6.63) depends on the choice of origin of y, and we have assumed that the origin is such that the relative phase is zero. It is easy to see that $\psi(x, y)$ describes a vortex state in the film. In fact $|\psi|^2$ has zeros along the line $x = 0$ at points separated in y by $\lambda_y = \pi/k$. In terms of x_0,

$$\lambda_y = \phi_0/\mu_0 H_0 x_0. \tag{6.64}$$

The level surfaces of $|\psi|^2$ are sketched in Fig. 6.17, and it can be seen that we do have a vortex state. As the reduced thickness $d/2\xi(T)$ decreases, x_0 decreases. This means that λ_y increases, so that as the film gets thinner it becomes progressively harder to squeeze vortices into the film, and the vortices become more elongated. Finally, for $d/\xi(T)$ less than the critical value, x_0 is zero: there is no degeneracy, and no vortex state. The film is simply too thin for the vortices to fit in.

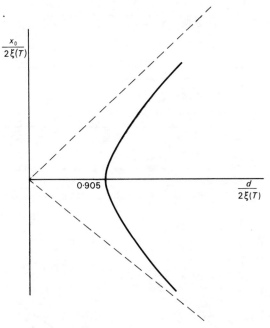

FIG. 6.15   To show the variation of the well centre x_0 with film thickness.

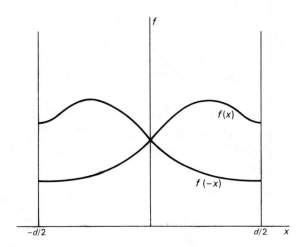

FIG. 6.16   The two degenerate wave functions $f(x)$ and $f(-x)$ for $x_0 \neq 0$.

Experimentally, the occurrence of a vortex state for films with $\varepsilon > \varepsilon_{cr}$ shows up quite clearly. The possibility was first appreciated by Sutton (1966). Fig. 6.18 shows the results of tunnelling measurements on an N–S junction, in which the S film is an In–Bi alloy. The measurement is of the zero-voltage conductance of the film $D_{V=0}$ against parallel magnetic field at various temperatures. At the lower temperatures, for which $\varepsilon > \varepsilon_{cr}$, there is a pronounced kink in the curve at the field

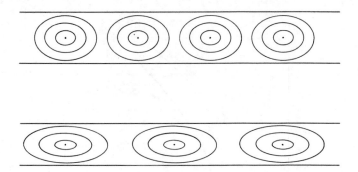

FIG. 6.17   Level surfaces of $|\psi|^2$ for two different values of $d/\xi(T)$, and hence of λ_y.

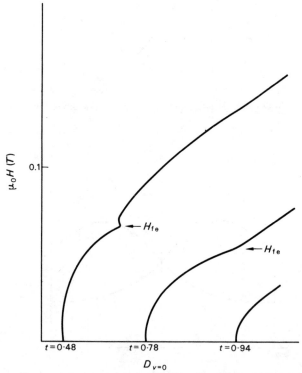

FIG. 6.18   Tunnelling conductance curves on an In alloy film 190 nm thick showing flux entry effect. (After Guyon *et al.*, 1967.)

H_{fe} for which the vortex lattice first enters the film. At a high reduced temperature with $\varepsilon < \varepsilon_{cr}$, no such anomaly is observed. The field H_{fe} is shown in Fig. 6.14, together with the critical fields. As expected, H_{fe} merges with $H_{c\parallel}$ at $\varepsilon = \varepsilon_{cr}$.

We have now given a fairly complete discussion of second-order critical fields for a thin film in a parallel field. It will be appreciated that it is also possible to have first-order transitions; for example a very thick film with a small value of κ will undergo a first-order transition at a critical field more or less equal to H_{cb}. Our discussion is therefore restricted as it stands to films with a sufficiently high κ-value, typically alloy films. In low κ films, the coherence length $\xi(T)$ is long, which means that the approximation $\psi = $ constant is often adequate. We shall discuss both first- and second-order critical fields of low κ films in § 6.6, using that approximation.

6.5 Abrikosov vortex lattice

In the last section we gave an account of the calculation of second-order critical fields with the use of the linearized GL equations. At H_{c2} and H_{c3}, ψ varies spatially, over the characteristic distance $\xi(T)$, whereas in calculating the critical field of a film of thickness $d < \xi(T)$, we were able to take ψ as a constant. We now go on to look at some consequences of the non-linear term $\beta|\psi|^2\psi$ in the GL equation. The simplest case to deal with, naturally, is a thin film in which ψ can be taken as a constant; as we shall see in the next section, one can then find expressions for ψ and the free energy for all values of the applied field H_0. First, in this section, we shall discuss the structure of the mixed state for H_0 just below H_{c2}. In the mixed state, ψ varies spatially, and it is not possible to deal analytically with the non-linear terms for all values of H_0. However, we are dealing with a second-order phase transition, and just below H_{c2} ψ is small (Fig. 6.5), so that we can handle the non-linear terms by a perturbation method. The solution of this problem is due to Abrikosov (1957), and we shall give an introductory account which follows Abrikosov's paper fairly closely. Because of the importance of the problem, a number of equivalent formulations of the solution have been given, for example by de Gennes (1966) and by Fetter and Hohenberg in Parks (1969).

We have to solve the two GL equations, (6.18) for ψ and (6.23) for the current **J**. We take the applied field H_0 to be just less than H_{c2}, so that the vector potential **A** satisfies

$$\text{curl } \mathbf{A} = \mathbf{B} = \mu_0 \mathbf{H}_0 + \mathbf{B}_{loc}, \tag{6.65}$$

where $\mathbf{B}_{loc}$ is the field generated by the supercurrents:

$$\text{curl } \mathbf{B}_{loc} = \mu_0 \mathbf{J}. \tag{6.66}$$

It is convenient to write

$$\mathbf{A} = \mathbf{A}_{c2} + \mathbf{A}_1, \tag{6.67}$$

where $\mathbf{A}_{c2}$ is the vector potential of the field H_{c2}, given explicitly by

$$\mathbf{A}_{c2} = (0, \mu_0 H_{c2} x, 0). \tag{6.68}$$

We take all magnetic fields to be directed along the z-axis.

Just below H_{c2}, we can expect ψ to be close to a solution of the linear equation, eq. (6.35), with H_{c2} in place of H_0. We therefore put

$$\psi = \psi_L + \psi_1, \tag{6.69}$$

where ψ_L is a solution of the linear equation. However, eq. (6.69) begs two questions about ψ_L. First, because we have a non-linear system, the normalization of ψ_L is important; eventually the normalization determines the strength of the supercurrent **J** and therefore the induction **B**. To make sure that we have the correct normalization, we shall require that ψ_L and ψ_1 are orthogonal,

$$\int \psi_L^* \psi_1 \, d^3\mathbf{r} = 0. \tag{6.70}$$

We can expand ψ_1 as a series in terms of the eigenfunctions of the oscillator Hamiltonian of eq. (6.35), and eq. (6.70) is the condition that the lowest eigenfunction ψ_L does not occur in ψ_1. The second question about ψ_L arises because the lowest eigenstate of eq. (6.35) is highly degenerate; the solution can have any values of k_y and x_0 related by eq. (6.40). Again, the non-linearity of the system lifts the degeneracy, and singles out one particular solution. At the outset, however, we simply choose a general linear combination of solutions with various x_0 values:

$$\psi_L(x, y) = \sum_n C_n \exp{(inky)} \exp{[-(x - x_n)^2/2\xi^2(T)]}, \tag{6.71}$$

with

$$x_n = nhk/2e\mu_0 H_{c2} = nk\xi^2(T). \tag{6.72}$$

It will be seen that eq. (6.71) is periodic in y with wavelength $2\pi/k$. Following the original calculation of Abrikosov, we shall consider only combinations such that $|\psi_L|^2$ is periodic in x as well as in y. We can ensure this with the periodicity condition

$$C_{n+N} = C_n. \tag{6.73}$$

In fact the only values of N that have been considered are $N = 1$ (Abrikosov, 1957) and $N = 2$ (Kleiner et al., 1964), the latter giving the triangular lattice which is observed in practice. The choice of a periodic solution in eq. (6.71) is nowadays rather obvious, since we know from the discussion of § 4.3 that the mixed state consists of a periodic array of vortices.

    Most of the analysis is independent of the detailed structure of ψ_L, that is, the particular choice of C_n and k in eq. (6.71). The first important result concerns the current $\mathbf{J}_L$ associated with ψ_L. If eq. (6.71) is substituted into eq. (6.23) for the current, it can be shown that

$$J_{Lx} = -\frac{eh}{m}\frac{\partial}{\partial y}|\psi_L|^2 \tag{6.74}$$

$$J_{Ly} = \frac{eh}{m}\frac{\partial}{\partial x}|\psi_L|^2. \tag{6.75}$$

This means, by comparison with eq. (6.66), that the field generated by $\mathbf{J}_L$, which is in the z-direction of course, is

$$B_{loc} = -\mu_0 eh|\psi_L|^2/m. \tag{6.76}$$

This gives us the important result that the lines of constant **B** coincide with the lines of constant $|\psi_L|^2$, and that these lines are also the lines of current flow, $\mathbf{J}_L$.

Thus the contours in diagrams like Figs. 6.19 and 6.20 represent at the same time level surfaces of $|\psi_L|^2$ and of $\mathbf{B}$, and streamlines of the current $\mathbf{J}_L$.

We can now look at the normalization of ψ_L. We split up $\mathbf{A}$ and ψ as in eqs. (6.67) and (6.69), and rewrite the GL equation for ψ as an equation for the small correction ψ_1:

$$\frac{1}{2m}(-i\hbar\nabla - 2e\mathbf{A}_{c2})^2\psi_1 + \alpha\psi_1 + \frac{1}{2m}(-i\hbar\nabla - 2e\mathbf{A}_{c2} - 2e\mathbf{A}_1)^2\psi_L$$
$$+ \alpha\psi_L + \beta|\psi_L|^2\psi_L = 0. \tag{6.77}$$

We treat A_1, ψ_1 and $\beta|\psi_L|^2\psi_L$ as small quantities, and retain only first-order terms. For convenience, we have kept in the zero'th-order term $\mathscr{H}_0\psi_L$, where

$$\mathscr{H}_0 = \frac{1}{2m}(-i\hbar\nabla - 2e\mathbf{A}_{c2})^2 + \alpha; \tag{6.78}$$

we have

$$\mathscr{H}_0\psi_L = 0, \tag{6.79}$$

since this is the defining equation for ψ_L. The term involving ψ_1 in eq. (6.77) is $\mathscr{H}_0\psi_1$. If we think again of ψ_1 as a sum of eigenfunctions of $\mathscr{H}_0$, not including ψ_L, then $\mathscr{H}_0\psi_1$ is also a sum of eigenfunctions not including ψ_L. Hence our normalization condition, eq. (6.70), is equivalent to

$$\int\psi_L^*\mathscr{H}_0\psi_1\,d^3\mathbf{r} = 0. \tag{6.80}$$

We can substitute for $\mathscr{H}_0\psi_1$ from eq. (6.77), to get the explicit form

$$\int\left\{\frac{1}{2m}\psi_L^*(-i\hbar\nabla - 2e\mathbf{A}_{c2} - 2e\mathbf{A}_1)^2\psi_L + \alpha|\psi_L|^2 + \beta|\psi_L|^4\right\}d^3\mathbf{r} = 0. \tag{6.81}$$

Equations (6.76) and (6.81) solve the problem for us, in principle, since eq. (6.76) gives $\mathbf{B}$ in terms of $|\psi_L|^2$, and eq. (6.81) determines $|\psi_L|^2$. In order to make the solution explicit, we have to reorganize eq. (6.81) somewhat. First, we integrate the first term by parts, to introduce $\nabla\psi_L^*$. The zero'th-order part of the first term cancels the integral of $\alpha|\psi_L|^2$, and the rest, ignoring the term in A_1^2, gives

$$\int(-\mathbf{A}_1\cdot\mathbf{J}_L + \beta|\psi|^4)\,d^3r = 0, \tag{6.82}$$

where $\mathbf{J}_L$, as before, is the current associated with ψ_L. With the use of eq. (6.66), the first term here is

$$-\int\mathbf{A}_1\cdot\mathbf{J}_L\,d^3\mathbf{r} = -\frac{1}{\mu_0}\int\mathbf{A}_1\cdot\text{curl}\,\mathbf{B}_{loc}\,d^3\mathbf{r}$$

$$= -\frac{1}{\mu_0}\int\mathbf{B}_{loc}\cdot\text{curl}\,\mathbf{A}_1\,d^3\mathbf{r}, \tag{6.83}$$

where again we have integrated by parts, using the vector identity $\text{div}\,(\mathbf{A}\times\mathbf{B}) = \mathbf{B}\cdot\text{curl}\,\mathbf{A} - \mathbf{A}\cdot\text{curl}\,\mathbf{B}$. From eqs. (6.65) and (6.67) we have

$$\text{curl}\,\mathbf{A}_1 = \mu_0(\mathbf{H}_0 - \mathbf{H}_{c2}) + \mathbf{B}_{loc}. \tag{6.84}$$

Putting eqs. (6.84) and (6.83) into eq. (6.82), and using eq. (6.76) for $\mathbf{B}_{loc}$, we find

$$(2\kappa^2 - 1)\frac{eh}{m}\langle|\psi_L|^4\rangle = (H_{c2} - H_0)\langle|\psi_L|^2\rangle, \tag{6.85}$$

where we have introduced the notation

$$\int |\psi_L|^2 \, d^3\mathbf{r} = \langle |\psi_L|^2 \rangle, \text{ etc.} \tag{6.86}$$

and replaced β by κ^2, with eq. (6.49).

Equation (6.85) is essentially as far as we can go without investigating the detailed form of the flux lattice described by ψ_L. It is convenient to summarize the properties of the lattice, following Abrikosov, in the parameter β_A:

$$\beta_A = \langle |\psi_L|^4 \rangle / (\langle |\psi_L|^2 \rangle)^2; \tag{6.87}$$

clearly we have $\beta_A \geqslant 1$ for any form of ψ_L. Equation (6.85) then yields

$$\langle |\psi|^2 \rangle = \frac{m}{e\hbar} \frac{H_{c2} - H_0}{(2\kappa^2 - 1)\beta_A}. \tag{6.88}$$

The average induction, from eqs. (6.65) and (6.76) is

$$\langle B \rangle = \mu_0 H_0 - \frac{\mu_0 (H_{c2} - H_0)}{(2\kappa^2 - 1)\beta_A} \tag{6.89}$$

and equivalently the magnetization is

$$M = - \frac{\mu_0 (H_{c2} - H_0)}{(2\kappa^2 - 1)\beta_A}. \tag{6.90}$$

It is easy to see from eq. (6.90) that the Gibbs energy G decreases as β_A decreases—the details are left for Problem 6.14. The choice of the parameters C_n and k in eq. (6.71) must therefore be made in such a way as to minimize β_A. Abrikosov originally chose all C_n equal, and $k = (2\pi)^{1/2} / \xi(t)$, which gives a square lattice, illustrated in Fig. 6.19, with $\beta_A = 1\cdot18$. Later, Kleiner et al. (1964) showed

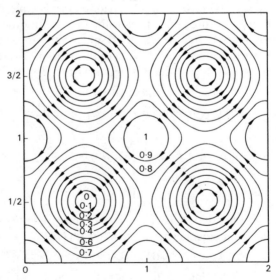

Fig. 6.19   Level surfaces of $|\psi_L|^2$ for the Abrikosov square lattice. The axes are marked in units of $(2\pi)^{1/2}\xi(T)$. (From Abrikosov, 1957.)

that the choice $N = 2$ in the periodicity eq. (6.73), together with

$$C_1 = \pm iC_0 \tag{6.91}$$

which corresponds to a triangular lattice, gives $\beta_A = 1.16$. Furthermore, the square lattice can be sheared continuously into the triangular lattice with β_A decreasing all the time, so that there is no question of metastability of the square lattice. The triangular lattice is shown in Fig. 6.20; it will be seen that it is the same as the lattice photographed by Essmann and Träuble (1967), Fig. 1.23.

FIG. 6.20   Level surfaces of $|\psi_L|^2$ for the triangular vortex lattice. The vertical distance between vortex cores is $2\pi^{1/2}\xi(T)/3^{1/4}$, and the horizontal distance between the rows of vortices is $3^{1/4}\pi^{1/2}\xi(T)$. (From Kleiner et al., 1964.)

Perhaps the most important feature of the results we have just derived is that the magnetization, eq. (6.90), depends upon the same parameter κ as gives the critical field H_{c2}. The GL equations, in the form in which we have stated them, are obviously valid only in some temperature interval near T_c, since we started with Landau's assumption that the free energy can be expanded as a power series in the order parameter. In fact, as we shall see in § 6.7, for alloys, in which the electrodynamics are local, the microscopic theory can be solved at all temperatures for $H \sim H_{c2}$, using a method based on the Abrikosov calculation. The result is that the equations for H_{c2} and the magnetization M continue to hold, except that the κ-parameters involved are different functions of temperature:

$$H_{c2} = \kappa_1(T)\sqrt{2}H_{cb} \tag{6.92}$$

$$M = -\frac{\mu_0(H_{c2} - H_0)}{\{2\kappa_2^2(T) - 1\}\beta_A}. \tag{6.93}$$

From this more recent point of view, one can say that the result of the GL calculation is that the κ-parameters coincide at $T = T_c$. Fig. 6.21 shows the temperature variations of κ_1 and κ_2 in pure Nb, and it can be seen that the limiting value at T_c is the same for both. The same is true for Figs. 6.35 and 6.36, which show κ_1 and κ_2 for a range of Nb alloys.

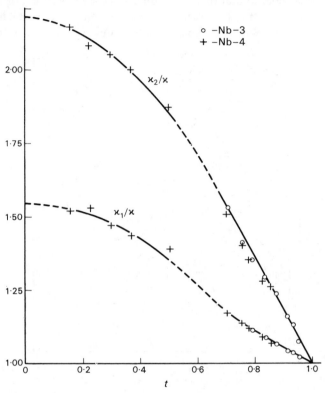

FIG. 6.21 κ_1/κ and κ_2/κ as functions of reduced temperature for two Nb specimens. κ is the GL value (limit at T_c), equal to 0·84 for both specimens. (After McConville and Serin, 1965).

6.6   Parallel-field phase transition in low κ films

As our final application of GL theory, we now discuss, following Ginzburg (1958), the behaviour of the Gibbs energy and the nature of the phase transition in a parallel field in a film of a low κ material, that is, one in which $\xi(T) \gg \lambda(T)$. As we have already pointed out, this is a fairly simple situation: for $d < \xi(T)$, which now covers an extended range of thickness, ψ cannot vary across the thickness of the film. Since $\lambda(T)$ is smaller than $\xi(T)$, there is a range of thickness (and temperature) in which the inequality $\lambda(T) < d < \xi(T)$ is satisfied, and in this range $\mathbf{B}$ and $\mathbf{A}$ can vary across the thickness of the film. However, because ψ is a constant, say ψ_0, we can find an explicit expression for the variation of $\mathbf{B}$ within the film, and this enables us to find the free energy G as a function of ψ_0 and the applied field H_0. Knowing G we can, of course, discuss the nature of the phase transition. In fact, we already know the limiting behaviour for small and large d. For $d < \lambda(T)$ as well as $d < \xi(T)$, both $\mathbf{B}$ and ψ are constant in the film. Our calculation of §6.4 therefore applies, and we have a second-order transition with the critical field given by eq. (6.58), or equivalently eq. (6.62). For large d, on the other hand, we have effectively a bulk type I superconductor, with a first-order transition at H_{cb}.

We again use the coordinate system of Fig. 6.13, with the z-axis along the applied field, and the x-axis normal to the film. We take the gauge

$$\mathbf{A} = \{0,\, A(x),\, 0\}, \tag{6.94}$$

so that the field $\mathbf{B}$ within the film is

$$\mathbf{B} = (0, 0, dA/dx). \tag{6.95}$$

It is convenient to express the constant value of ψ in terms of the bulk equilibrium value:

$$\psi^2 = f^2 |\alpha|/\beta. \tag{6.96}$$

The London equation then becomes, in place of eq. (6.27),

$$\mu_0 \mathbf{J}_e = -f^2 \mathbf{A}/\lambda^2, \tag{6.97}$$

where λ is the bulk penetration depth. We combine eq. (6.97) in the usual way with the Maxwell equation curl $\mathbf{B} = \mu_0 \mathbf{J}_e$, to get

$$\frac{d^2 A}{dx^2} = \frac{f^2 A}{\lambda^2}. \tag{6.98}$$

Our boundary condition is that $B = \mu_0 H$ at the boundaries of the film, so that the appropriate solution of eq. (6.98) is

$$A = \frac{\mu_0 H_0 \lambda}{f} \frac{\sinh(fx/\lambda)}{\cosh(fd/2\lambda)} \tag{6.99}$$

$$B = \mu_0 H_0 \frac{\cosh(fx/\lambda)}{\cosh(fd/2\lambda)}. \tag{6.100}$$

These equations for the field quantities contain the result that if ψ^2 is decreased by a factor f, λ is increased by the factor $1/f$. This is an obvious consequence of the reduction in the strength of the screening currents.

We now have simply to evaluate the Gibbs free energy, given by eqs. (6.15) and (6.16). With ψ constant, the free energy becomes

$$G_s - G_n = S \int \left\{ \alpha |\psi|^2 + \frac{1}{2}\beta|\psi|^4 + \frac{2e^2}{m} A^2 |\psi|^2 + \frac{1}{2\mu_0} (B - \mu_0 H_0)^2 \right\} dx, \tag{6.101}$$

where S is the surface area of the film. Substituting in eq. (6.96) for ψ, and eqs. (6.99) and (6.100) for the field quantities, we find

$$G = \frac{G_s - G_n}{V|\alpha|^2/2\beta}$$

$$= f^4 - 2f^2 + \left(\frac{H_0}{H_{cb}}\right)^2 \left(1 - \frac{2\lambda}{fd}\tanh\frac{fd}{2\lambda}\right). \tag{6.102}$$

where $V = Sd$ is the volume of the film, and we have used eq. (6.32) to introduce the bulk critical field H_{cb}.

To find the Gibbs energy G, which is a function of T and H_0, we have to minimize eq. (6.102) with respect to f. It is instructive to begin by considering G as a function of f for various fixed values of H_0/H_{cb}. In general, dG/df is zero at $f = 0$, since G contains only even powers of f; $f = 0$ (the normal state) therefore always gives a local minimum or a local maximum of G. For large f, G increases, since the term in f^4 is then dominant. For a small enough field H_0/H_{cb}, we expect the superconducting phase to be stable, so that $f = 0$ is a maximum, and there is a minimum at some non-zero value of f. For a sufficiently large H_0/H_{cb}, conversely, the normal phase must be stable, so that $f = 0$ is a minimum and there is no other minimum. This expected behaviour is shown in Fig. 6.22.

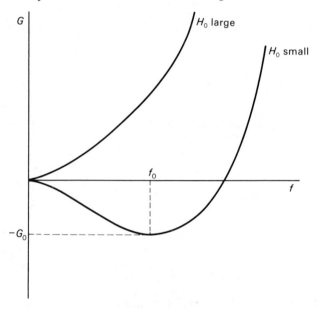

FIG. 6.22   Expected behaviour of G for small and large H_0. When the superconducting phase is stable, it occurs at the value f_0 of f which gives a minimum in G, and the value of G at the minimum gives the condensation energy G_0.

We have already seen that we may expect to find a second-order phase transition in a sufficiently thin film, and a first-order transition in a sufficiently thick one. In fact d enters G only in the reduced form d/λ, so that we may expect to find a critical value d_{cr}/λ below which the transition is second order, and above which it is first order. We can see how to find d_{cr}/λ if we look a little more closely at the variation of G with f. With a first-order transition, the situation is somewhat analogous to that sketched for a bulk specimen in Fig. 6.7. As we show in Fig. 6.23, as the field H_0 is increased from zero, the minimum which corresponds to the superconducting phase moves up in energy, and in fact the equilibrium value f_0

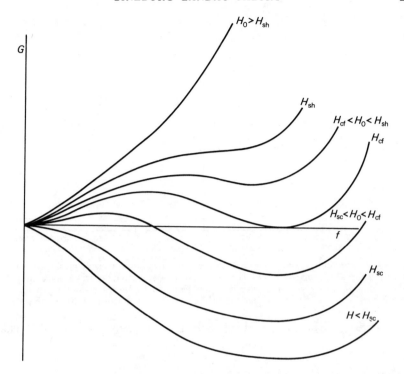

FIG. 6.23 Variation of G with f at various fields in the case of a first-order phase transition.

decreases. At the thermodynamic critical field H_{cf} both the normal phase, with $f = 0$, and the superconducting phase, with $f = f_0 \neq 0$, are stable and correspond to local minima; the energies of the two phases are of course equal. As always with a first-order transition, there is the possibility of *supercooling* of the normal phase and *superheating* of the superconducting phase. That is, the normal phase is metastable in a field range $H_{sc} < H < H_{cf}$, and the superconducting phase is metastable in a range $H_{cf} < H < H_{sh}$. At the maximum supercooling field H_{sc} the local minimum at $f = 0$ turns into a local maximum, and a similar property identifies the maximum superheating field H_{sh}. The corresponding G–f curves are shown in Fig. 6.23. It is also of interest to look at the variation with H_0 of f_0, the value of f which gives the superconducting phase. This is shown in Fig. 6.24.

We can readily derive equations for the various critical fields. For a given H_0, the value f_0 which corresponds to the superconducting phase is given by

$$\partial G/\partial f = 0, \tag{6.103}$$

which can be written as

$$\left(\frac{H_0}{H_{cb}} \right)^2 = \frac{4f_0^2(f_0^2 - 1)\cosh^2\left(f_0 d/2\lambda\right)}{1 - (\lambda/f_0 d)\sinh\left(f_0 d/\lambda\right)}. \tag{6.104}$$

At H_{cf}, eq. (6.103) determines f_0, and then H_{cf} is given by the additional condition

$$G = 0. \tag{6.105}$$

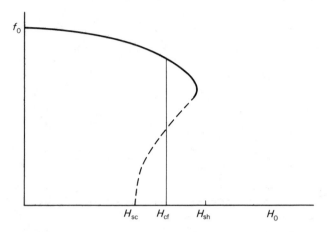

FIG. 6.24   Variation of f_0 with H_0 for a first-order phase transition. The full line is the position of the local minimum which corresponds to the superconducting phase, and the broken line is the position of the maximum which separates the two possible phases for fields between H_{sc} and H_{sh}. In the absence of supercooling or superheating, the phase transition occurs with a discontinuity in f at the field H_{cf}.

The superheating field is that at which the extremum given by eq. (6.104) changes from a minimum to a maximum, so the additional condition is

$$\frac{\partial^2 G}{\partial f^2} = 0 \qquad \text{at} \qquad f = f_0. \tag{6.106}$$

Finally, the supercooling field is given by

$$\frac{\partial^2 G}{\partial f^2} = 0 \qquad \text{at} \qquad f = 0. \tag{6.107}$$

This last condition is easy to make explicit, since it corresponds to equating to zero the coefficient of f^2 in a power-series expansion of G. The result is

$$H_{sc}/H_{cb} = 2\sqrt{6}\lambda/d. \tag{6.108}$$

In fact, if we express H_{cb} and λ in terms of the GL parameters $|\alpha|$ and β, we can see that this is the same as the second-order critical field H_c of eq. (6.58), that is, it coincides with the parabolic part of the critical field curve of Fig. 6.14. This is to be expected, since in both cases we derived the critical field on the assumption that ψ is uniform across the film and that we are dealing with a transition in which $\psi \to 0$.

When the phase transition is of second order, things are naturally rather simpler. The variation of G with f for different field values is then as shown in Fig. 6.25: as H_0 increases, the minimum corresponding to the superconducting phase moves in, and reaches $f_0 = 0$ at the critical field H_c. The variation of f_0 with field is shown in Fig. 6.26. There is, of course, no possibility of superheating or supercool-

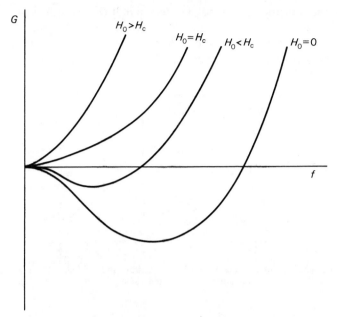

FIG. 6.25 Variation of G with f at various fields for a second-order phase transition.

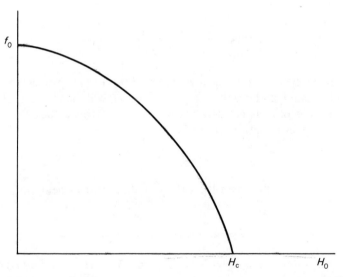

FIG. 6.26 Variation of f_0 with H_0 for a second-order phase transition.

ing. At the critical field H_c, the maximum in G at $f = 0$ just coincides with the minimum at f_0 as $f_0 \to 0$, so H_c continues to be given by eq. (6.107), with the explicit expression of eq. (6.108).

We can now see how to find the critical value d_{cr}/λ at which the nature of the phase transition changes. Consider an applied field H_0 slightly lower than H_{sc} of eq. (6.108). If the transition is of second order, so that H_{sc} is the critical field, then there is a minimum in G, corresponding to the superconducting phase, at a small value of f. On the other hand, if the transition is of first order and H_{sc} is the supercooling field, then there is no minimum for small f. These two possibilities are illustrated in Fig. 6.27.

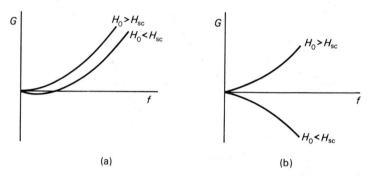

FIG. 6.27   Variation of G at small values of f for fields just above and just below H_{sc}. (a) second-order transition; (b) first-order transition.

Suppose we expand the condition for a minimum, eq. (6.104), to second order in f_0:

$$\left(\frac{H_0}{H_{cb}}\right)^2 - \frac{24\lambda^2}{d^2} = f_0^2 \frac{24\lambda^2}{d^2}\left(\frac{d^2}{5\lambda^2} - 1\right). \tag{6.109}$$

The vanishing of the left-hand side gives the field H_{sc}. For H_0 just less than H_{sc} the left-hand side is small and negative. Equation (6.109) will then have a solution for f_0 if the coefficient of f_0^2 is negative. As we have seen, this is the condition for a second-order phase transition, so we have

$$d_{cr} = \sqrt{5}\lambda. \tag{6.110}$$

It is useful also to write down the corresponding critical field value from eq. (6.108):

$$H_{cr} = 2\sqrt{6}H_{cb}/\sqrt{5} = 2\cdot19 H_{cb}. \tag{6.111}$$

The phase transition is first order for $d > d_{cr}$ and $H_{cf} < H_{cr}$, and second order for $d < d_{cr}$ and $H_{cf} > H_{cr}$.

The above account summarizes the main predictions for films given by Ginzburg (1958). In addition, Ginzburg gave the corresponding results for a sphere and for a cylinder parallel to the applied field. Experimentally, a first-order transition can be identified as one for which hysteresis is observed. Thus Fig. 6.28 shows traces of the resistance in increasing and decreasing field for a thick film in a parallel field. The appearance or otherwise of hysteresis is in good agreement with

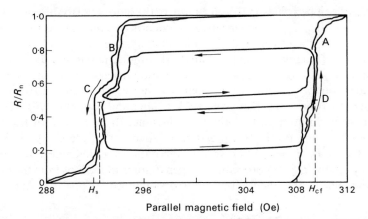

FIG. 6.28 Parallel-field resistive transition of a tin film of thickness 465 nm at 1·65 K. The minor hysteresis loops are generated by reversing the field sweep at an intermediate point like C or D. (After Baldwin, 1963.)

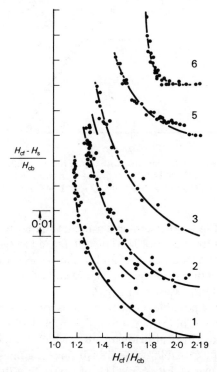

FIG. 6.29 Extent of hysteresis in measurements of resistive transition. (From Baldwin, 1964.)

eq. (6.111), as can be seen from Fig. 6.29, where the amount of hysteresis is plotted as a function of H_{cf}/H_{cb}, which means as a function of temperature, for a number of specimens. On the other hand, the amount of hysteresis, $H_{cf} - H_s$, is generally up to an order of magnitude smaller than the theoretical maximum $H_{sh} - H_{sc}$. This is no doubt because of the presence of nucleating defects within the film.

The variation of f_0 with H_0 as sketched in Figs. 6.24 and 6.26 has been investigated by tunnelling (Douglass, 1961, Collier and Kamper, 1965). The raw data consist of variation of tunnelling conductance dI/dV, say, with H_0. This is interpreted by using the BCS expression for dI/dV with the energy gap $\Delta(T)$ replaced by $\Delta(H_0, T) = f_0 \Delta(T)$, so that the data are converted into measurements of f_0 versus H_0. This procedure is open to some objections from the point of view of the microscopic theory, but the errors involved are probably not serious.

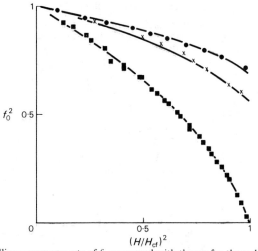

FIG. 6.30   Tunnelling measurements of f_0 compared with theory for three different specimens. Data from Collier and Kamper (1966). The theoretical curves are taken from eq. (6.104), with the parameter d/λ determined by the critical field value. (From Tilley et al., 1966.)

An example is shown in Fig. 6.30, where experimental data are compared with theoretical curves derived from eq. (6.104). The difference between the first- and second-order transitions shows up clearly, and it can be seen that the fit of the theoretical curves to the data is very good. The results of a similar analysis of ultrasonic attenuation data are shown in Fig. 6.31.

To get a full picture of the possible phase transitions for films in a parallel field we have to combine the results of this section, which apply to low κ films, with the calculations of second-order critical fields for all κ which we gave in §6.4. It is convenient to represent the possible phase transitions on a phase diagram with axes κ and $d/\lambda(T)$, as shown in Fig. 6.32. The use of such a diagram was first suggested by Arp et al. (1966), and a corrected version of their diagram was given by Guyon et al. (1967). The result of this section is that for small κ, the phase transition is first order for $d/\lambda > \sqrt{5}$. In the bulk limit of large d/λ, the phase

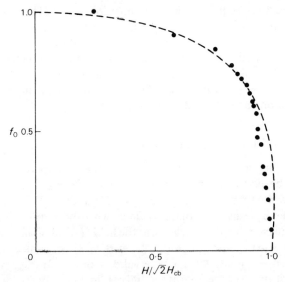

FIG. 6.31 f_0 versus H_0 deduced from ultrasonic surface wave attenuation on a Pb film 125 nm thick at $T = 5\,\mathrm{K}$. (After Jain and Mackinnon, 1970.)

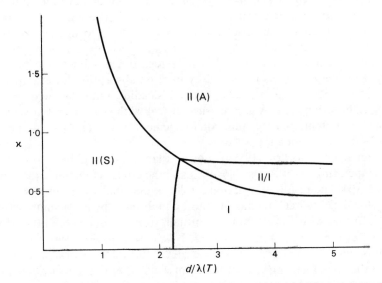

FIG. 6.32   Phase diagram for critical fields of a thin film in a parallel field. (Based on the diagram in Guyon *et al.*, 1967.)

transition is of first order for $\kappa < 0\cdot42 = \sqrt{2}/1\cdot69$. For $0\cdot42 < \kappa < 0\cdot71$ in the bulk limit we have $H_{c2} < H_{cb} < H_{c3}$; this means that as the field is lowered there is first a second-order transition to the sheath state, then a first-order transition at H_{cb}. This is shown as region II/I. For $\kappa > 0\cdot71$ in the bulk limit we have $H_{cb} < H_{c2} < H_{c3}$; accordingly there is a second-order transition to the sheath state at H_{c3}, followed by a second-order bulk transition at H_{c2}. This is shown as region II(A). Finally, we saw in § 6.4 that the character of the wave function below the second-order transition depends upon $\varepsilon = d^2/4\xi^2(T)$: for $\varepsilon < \varepsilon_{cr} = 0.82$ we have a symmetric wave function, while for $\varepsilon > \varepsilon_{cr}$ we have an asymmetric wave function and the actual state is the squashed vortex array of Fig. 6.17. The curve $\varepsilon = \varepsilon_{cr}$ appears on the phase diagram as $d/\lambda(T) = 1.8/\kappa$, and we label the symmetric and asymmetric regions as II(S) and II(A) respectively.

6.7 Relation of GL theory to microscopic theory

In the previous sections we have applied the GL theory to a number of problems. We based the calculations solely on the form of the free energy which was originally postulated by Ginzburg and Landau (1950) well before the microscopic theory was developed. In fact, with the notable exception of the prediction of surface sheath nucleation at the field H_{c3}, all of the results we derived were already known before the date of the BCS paper. Once the microscopic theory had been formulated, the whole validity of the GL theory was naturally called into question — see the note at the end of Ginzburg (1958), for example. In its original formulation, the BCS theory could not easily be extended to situations in which an applied magnetic field induced spatial variations of the order parameter. Gorkov (1958), however, reformulated the BCS theory in the language of Green's functions, and showed that one could then allow fairly easily for spatial variations. A bulk superconductor, as we know, is characterized by an energy gap Δ. Gorkov's first result is that in general we must regard Δ as a function of position, $\Delta = \Delta(\mathbf{r})$. The energy gap, as measured by infrared absorption for example, is a property of the specimen as a whole, so it is misleading and in fact wrong to call $\Delta(\mathbf{r})$ a spatially varying energy gap; it is usually called the pair potential, since it is analogous to a potential-energy function for elementary excitations. This point is explained fully by de Gennes (1966).

In a later paper, Gorkov (1959a) showed that in the neighbourhood of the critical temperature T_c, it is possible to expand the Green's function equations in powers of $\Delta(\mathbf{r})$. The resulting equation for $\Delta(\mathbf{r})$ is essentially the same as the GL equation for $\psi(\mathbf{r})$: in fact $\Delta(\mathbf{r})$ has only to be multiplied by a constant factor to bring it into line with the customary normalization of ψ (we recall that this is determined by the choice of the coefficient $1/2m$ for the gradient term). Likewise, the equation for the supercurrent has the usual GL form. Thus Gorkov was able to show that there is some temperature range for which the GL theory is valid. Later work has been devoted to exploring the ranges of validity of GL theory, generally using the Gorkov formalism, and we shall presently summarize the main features. It is of course the fact that GL results are derived, at one remove, from the microscopic theory that ensures that GL theory is of continuing importance. The Gorkov formalism is powerful and elegant, but it does demand some knowledge of many-body theory. Good introductions to the formalism are given in Abrikosov et al. (1963) and in the chapter by Ambegaokar in Parks (1969).

The ranges of validity are mainly evident from the structure of the GL equations themselves. They obviously require two conditions:

A $\Delta(\mathbf{r})$ small

B   local electrodynamics.

In the light of the discussion of §3.4, leading up to eq. (3.34), we may rewrite condition B as

$$B \quad \xi_0 \ll \lambda(T) \qquad \text{pure metal} \tag{6.112}$$
$$l \ll \lambda(T) \qquad \text{alloy with } l \ll \xi_0 \tag{6.113}$$

In Gorkov's first paper (1959a), on pure superconductors, the condition $T_c - T \ll T_c$ was imposed to ensure that conditions A and B were satisfied. In fact condition B is very restrictive in a pure metal; the appropriate temperature interval for Sn, for example, is only about 0·1 K (Ginzburg, 1958). The situation in alloys is far more favourable; as we have seen before, the London condition of eq. (6.113) can be satisfied at all temperatures in alloys. In Gorkov's paper (1959b) on alloys, therefore, the condition $T_c - T \ll T_c$ was required only to meet condition A.

It was later realized by de Gennes (1964) and by Maki (1964), that in alloys condition A is satisfied in the vicinity of a second-order critical field like H_{c2}. Since condition B is not restrictive in alloys, one can derive a form of the GL theory near to H_{c2}; the only difference is that the parameters $|\alpha|$ and β no longer have the simple temperature dependences of the Landau theory, eqs. (6.6) and (6.7). The simplest situation is when $l \ll \xi_0$, which is called the dirty limit. The critical field H_{c2} is determined solely by $|\alpha|$, eq. (6.43). The theoretical form of the temperature variation of H_{c2} in the dirty limit is shown in Fig. 6.33, together with some experimental results. The extension of the linear GL theory to all temperatures

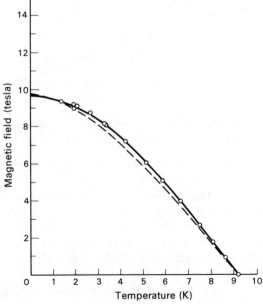

FIG. 6.33   Upper critical field of a Nb–37 at % Ti alloy, compared with dirty limit theory (dashed line). (After Shapira and Neuringer, 1965.)

means that all the results of the linear theory given in §6.4 continue to hold at all temperatures for alloys. In particular, we have $H_{c3} = 1{\cdot}69\,H_{c2}$, and the 'universal curve' of Fig. 6.14 for the critical fields of alloy films is valid, at all temperatures. In fact, the experimental results of Fig. 6.14 were obtained over an extended temperature range. It is customary to express the generalization of the linear theory in terms of a parameter $\kappa_1(T)$, as in eq. (6.92). The field H_{cb} there is taken to have the Gorter–Casimir temperature dependence $H_{cb} \propto 1 - t^2$. The temperature variation of H_{c2} shown in Fig. 6.33 translates into a variation of $\kappa_1(T)/\kappa$ from 1 at T_c to $1{\cdot}20$ at $T = 0$.

In addition to the linear term Maki (1964) also found the coefficient of the term in $|\Delta(\mathbf{r})|^2\Delta(\mathbf{r})$, that is to say the parameter β, at all temperatures in the dirty limit; his result was later corrected by Caroli et al. (1966). The Abrikosov analysis goes through unchanged, so that the conclusions reached in §6.5 hold at all temperatures in the dirty limit. The only modification, which we have already mentioned, is that since the κ_2-parameter, which determines the slope of the magnetization, eq. (6.93), is simply proportional to $\beta^{1/2}$, it has a slightly different temperature dependence from $\kappa_1(T) = H_{c2}/\sqrt{2}H_{cb}$. The predicted forms are shown in Fig. 6.34.

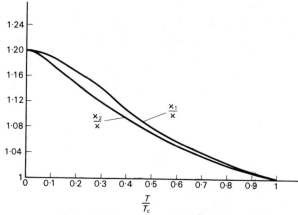

FIG. 6.34   Predicted temperature dependence of κ_1 (related to H_{c2}) and κ_2 (related to magnetization) in the dirty limit $l \ll \xi_0$. (From Caroli et al., 1966.)

The above account summarizes the main results for the dirty limit, which have been extensively applied to experiments on alloys. We can state the domains of validity as follows. In the neighbourhood of T_c, the original GL theory holds for all fields; nowadays people often write of the GLAG theory, because of the important contributions of Abrikosov and Gorkov. At all temperatures, the Maki–de Gennes generalization of the GL theory holds for fields H near H_{c2} ($H_{c2} - H \ll H_{c2}$), with the temperature dependence of the κ-parameters shown in Fig. 6.34.

Away from the dirty limit $l \ll \xi_0$, matters become more complicated. There are two kinds of complication. As long as the electrodynamics are still local, the GL equations hold for small $\Delta(\mathbf{r})$, but the coefficients α and β change because some

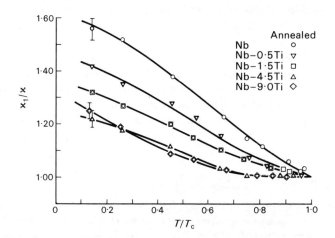

FIG. 6.35   The normalized function $\kappa_1(t)/\kappa$ for a range of annealed Nb alloy specimens. The value of the normalizing parameter κ is the limiting value of κ_1 as $t \to 1$. The data for this figure and Fig. 6.36 were obtained from magnetization measurements. (After Fietz and Webb, 1967.)

simplifications that are made in the dirty limit are no longer possible. Secondly, in a pure type II superconductor (Nb or V), the electrodynamics are not local over most of the temperature range, so no form of the GL theory holds. However, the calculations from the microscopic theory still give values of H_{c2} and of the magnetization near H_{c2}, and it is customary to convert these into parameters $\kappa_1(T)$ and $\kappa_2(T)$ via eqs. (6.92) and (6.93). The most extensive calculations of

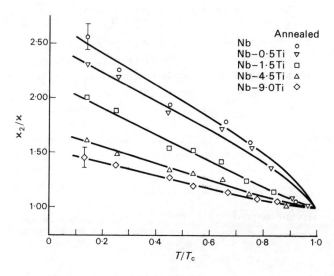

FIG. 6.36   The normalized function $\kappa_2(t)/\kappa$ for the specimens of Fig. 6.35. The limiting value as $t \to 1$ of κ_2 coincides with the limiting value of κ_1. (After Fietz and Webb, 1967.)

the temperature dependence of κ_1 and κ_2 are those of Eilenberger (1967). His general result is that both κ_1 and κ_2 are expected to increase more rapidly than in the dirty limit as T decreases, and further that $\kappa_2(T) \geqslant \kappa_1(T)$ at all temperatures. The exact form, however, depends on the ratio l/ξ_0 and also on the degree of anisotropy in the scattering of electrons off the individual impurity centres present. Some results from measurements on a range of Nb alloys are shown in Figs. 6.35 and 6.36. The experimental results are in qualitative agreement with the theory, although generally both κ-parameters increase more rapidly than theory indicates. Further details are given in the chapter by Serin in Parks (1969).

6.8   Ginzburg–Pitaevskii equations for Helium II

The Landau theory of second-order phase transitions was applied to liquid helium near the λ-point by Ginzburg and Pitaevskii (1958), henceforth referred to as GP. As in superconductors, the macroscopic wave function ψ is used as the order parameter for the transition. The normalization of ψ is chosen so that the superfluid density can be expressed as

$$\rho_s = m_4 \, |\psi(\mathbf{r})|^2 \tag{6.114}$$

which, of course, is the choice we made in Chapters 1 and 2 of this book. The derivation of the GP equation for ψ is parallel to the derivation of the GL equation (6.18), described in §6.3, but without the complications caused by the magnetic properties of superconductors.

GP treat the situation in which the normal fluid is at rest, and assume that the free-energy density can be expanded in powers of $|\psi|^2$ and $|\nabla\psi|^2$,

$$f(\mathbf{r}) = f_1 + \alpha |\psi(\mathbf{r})|^2 + \frac{1}{2}\beta |\psi(\mathbf{r})|^4 + \frac{\hbar^2}{2m_4} |\nabla\psi(\mathbf{r})|^2, \tag{6.115}$$

where f_1 is the free-energy density of He I. We note that it makes no essential difference whether $f(\mathbf{r})$ is the density of the Helmholtz free energy or the Gibbs energy. To find the stable state at a given temperature and pressure, the total free energy

$$F = \int f(\mathbf{r})\,\mathrm{d}^3\mathbf{r} \tag{6.116}$$

must be minimized with respect to the wave function. For minimization with respect to ψ^* the appropriate Euler–Lagrange equation has the form of eq. (6.17),

$$\frac{\partial f}{\partial \psi^*} - \sum_j \frac{\partial}{\partial x_j} \frac{\partial f}{\partial(\nabla_j \psi^*)} = 0. \tag{6.117}$$

When eq. (6.115) is substituted into eq. (6.117), GP obtain the following equation:

$$-\frac{\hbar^2}{2m_4} \nabla^2 \psi + \alpha\psi + \beta |\psi|^2 \psi = 0, \tag{6.118}$$

which is identical to the first GL equation (6.18) with the vector potential $\mathbf{A}$ put equal to zero.

Following the Landau theory, GP assume that the coefficients α and β in the free-energy expansion can themselves be expanded in powers of $(T - T_\lambda)$; in particular that

$$\alpha = A(T - T_\lambda) \tag{6.119}$$

where $A > 0$. Thus, for $T < T_\lambda$, $\alpha < 0$. In the same manner as for a superconductor, a coherence length for He II can be introduced into eq. (6.118):

$$\xi(T) = \hbar^2/2m_4|\alpha|. \tag{6.120}$$

GP estimate a value for the constant A in eq. (6.119), using data on the variation of specific heat on either side of T_λ, together with the gradient of the ρ_s versus T curve at T_λ. Combination of eq. (6.119) with eq. (6.120) then gives

$$\xi(T) \approx 4 \times 10^{-10}(T_\lambda - T)^{-1/2} \text{ m}. \tag{6.121}$$

The coherence length in He II defines the minimum range over which there will be appreciable variation in $|\psi|$ (or ρ_s). In other words, ξ should be the same as the *healing length* or the vortex core radius a_0, which we introduced in Chapter 4. We observe that $\xi(0)$ is of the same order as the value deduced for a_0 from the vortex-ring experiments (§4.2.5). Thus, except very close to the λ-point, $\xi(T)$ is comparable in magnitude with the interatomic spacing, a situation in which the Landau theory breaks down, because it is not then possible to expand the free energy as a power series with only a small number of terms. The coherence length exceeds the interatomic spacing by an order of magnitude or more when $(T_\lambda - T) < 10^{-2}$ K, but in this temperature range, fluctuations are dominant (see §6.10) and the Landau theory again breaks down. It seems therefore that the GP equation (6.118) is not valid for He II at any temperature.

In spite of these considerations, application of the GP theory to certain problems yields results which are supported, at least qualitatively, by experiments performed at temperatures well below T_λ. For example, GP use eq. (6.118) for a He II film, assuming the boundary conditions that ψ vanishes at a solid wall and at a free surface (cf. Fig. 2.12). At a fixed temperature there is a minimum film thickness $d_m = \pi\xi(T)$ below which there is no solution with $|\psi| \neq 0$, that is, for thickness smaller than d_m, $\rho_s = 0$. Alternatively, this result indicates that the superflow-onset temperature is depressed below T_λ in a thin film or narrow channel (Problem 6.19), behaviour that is well known from experiments (see Figs. 2.5, 2.10, 2.11). GP also use eq. (6.118) to find the variation of $|\psi|$ near a vortex.

To treat the case of superfluid flow through a channel, Mamaladze and Cheishvili (1966) have modified eq. (6.118) to include a term dependent on $(\nabla S)^2$, where S is the phase of the order parameter, and therefore on v_s^2. Their treatment indicates the possibility of observing the d.c. Josephson effect in He II (§5.5).

Pitaevskii (1958) generalized the basic GP theory to include the case when the normal fluid is not stationary, by adding the kinetic energy of the normal fluid to the total energy. In that case the free-energy density becomes

$$f'(\mathbf{r}) = \frac{1}{2}m_4 \left| \left(-\frac{i\hbar}{m_4}\nabla - \mathbf{v}_n\right)\psi \right|^2 + f_0(|\psi|^2), \tag{6.122}$$

where f_0 is the free-energy density of the stationary fluid in equilibrium. Khalatni-kov (1969) has discussed the minimization of $f'(\mathbf{r})$ and used it to obtain modified two-fluid equations which should be valid close to the λ-point. From these, he has derived formulae describing the dispersion of first sound and the attenuation of second sound near T_λ, finding good agreement with experimental results.

6.9   Broken symmetry

In this section and the next one we wish to take a general look at some similarities between the phase changes to superconductivity and superfluidity and other types of phase change. In this section we focus attention on the phase S of the superfluid wave function ψ. Strictly speaking, this section is out of place in a chapter on GL theory, because the comments we shall make are general and are not restricted to the temperature region in which the GL theory holds. However, it is convenient to use the GL theory as an example, to make the discussion rather more concrete.

We start with the observation that in an isolated specimen (of He II or a superconductor) the phase S has the same value throughout the specimen, but the free energy is independent of the value of S. This can be seen explicitly in the GL theory, where eq. (6.10) for the free energy is independent of S. Thus there is not just one equilibrium state, but an ensemble with all possible values of S. States in which S varies spatially, for example the mixed state, are not an exception, since the replacement $S(r) \to S(r) + S_0$, where S_0 is a constant, leaves the free energy unaltered. Here the ensemble of possible states involves all values of S_0. The same sort of situation holds for the ordinary Landau theory: it can be seen from Fig. 6.4 that $+\phi$ and $-\phi$ give the same free energy. We shall now see that a situation of this kind is general for all transitions into an ordered phase.

Consider, as we did at the beginning of §6.2, a model ferromagnetic system. We take a lattice of electron spins on fixed sites, each coupled to its neighbours by exchange forces. In a real magnetic system, the exchange force between two spins is a more or less complicated function of the distance between the spins and their relative orientation. We shall simplify matters by considering only interactions between nearest neighbours on the lattice of sites, and by assuming that the interaction is so anisotropic that it only couples the z-components of the spins. What we have is called the Ising model, and we may write the Hamiltonian explicitly as

$$\mathscr{H} = -J \sum_{\langle i,\, j \rangle} \sigma_z^i \sigma_z^j, \tag{6.123}$$

where the sum is restricted to nearest neighbours. With the interaction of eq. (6.123) it is energetically favourable for neighbouring spins to point in the same direction, so in the ground state at $T = 0$ the spins are all aligned. However, just as in the Landau theory, there is not one possible ground state, but two: the spins can point either up or down. The individual ground state, say with spins up, is described as a state of *broken symmetry*, because the ground state lacks a symmetry which is possessed by the Hamiltonian $\mathscr{H}$. To be precise, if the direction of the

z-axis is reversed, $\mathscr{H}$ is unaltered, whereas the spin-up ground state is transformed into the spin-down state. Thus the ensemble (spin-up + spin-down) retains the full symmetry of the Hamiltonian.

As the temperature is increased from zero, some of the spins reverse in direction, so that the average spin $\langle \sigma_z \rangle$ decreases. $\langle \sigma_z \rangle$ is proportional to the spontaneous magnetization, and it finally vanishes at the Curie temperature T_c, as we have already shown in Fig. 6.1. Above T_c, the magnet is in the paramagnetic phase, which has zero spontaneous magnetization. The paramagnetic phase does not have a broken symmetry: it is unaltered when the direction of the z-axis is reversed, so that it has the full symmetry of the Hamiltonian. We may therefore regard the phase change as T is decreased through T_c as going from a disordered phase into an ordered phase with a characteristic broken symmetry.

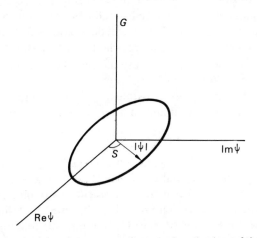

FIG. 6.37   The free energy is the same for all values of the phase S.

We now return to the phase change to superfluidity. We have already remarked that the free energy of the superfluid phase is independent of the phase S, and that we have an ensemble of different states corresponding to different values of S, as illustrated in Fig. 6.37. It is to be expected that a state with one particular value of S is a state of broken symmetry, so we have to see which is the symmetry which is possessed by the Hamiltonian but lacking in the state with fixed S. We recall from eq. (3.22) that in a superconductor changing S corresponds to a gauge transformation. We may likewise call a change of S in He II a gauge transformation. The Hamiltonian is obviously unaffected by a gauge transformation, whereas a state with fixed S is not. Thus the broken symmetry is the rather abstract one of *gauge symmetry*. Finally we note that the ensemble of states with all possible values of S is unaffected by a gauge transformation, that is, it is gauge symmetric like the Hamiltonian.

6.10   Fluctuation effects

We have already commented briefly, in discussing Fig. 6.3, that a Landau-type phase change is very sensitive to fluctuations. We now expand that comment. We defined the equilibrium value ϕ_0 of the order parameter ϕ as the value which minimizes the free energy F. We recall from elementary statistical mechanics that the probability of finding the system with a given value of ϕ is proportional to $\exp\{-F(\phi)/k_B T\}$. Thus ϕ_0 is not the only possible value of ϕ, but simply the *most probable* value. Thermal fluctuations have the effect that ϕ varies around ϕ_0, and the Landau theory makes the simplification of ignoring these variations. As we saw in Figs. 6.1 and 6.3, in the neighbourhood of T_c a large change in ϕ corresponds to a small change in F, so we may expect that near enough to T_c the fluctuations in ϕ are substantial and the Landau theory loses its validity.

We may use an argument first advanced by Ginzburg (1960) to estimate the extent of the temperature region near T_c in which fluctuation effects are dominant. A fluctuation into the superfluid state requires that the order parameter ψ increase from the normal state value of zero. Because the characteristic length for variations of ψ is $\xi(T)$, the fluctuation therefore occupies a minimum volume (in a bulk specimen) of order $\xi^3(T)$. The energy available to drive the fluctuation is of order $k_B T$, so fluctuation effects are dominant in a temperature region given by

$$k_B T > (F_n - F_s)\xi^3. \tag{6.124}$$

Since $F_n - F_s$ is proportional to $(T - T_c)^2$, and ξ^3 to $(T - T_c)^{-3/2}$, the right-hand side of this inequality is proportional to $(T - T_c)^{1/2}$.

It can be seen from eq. (6.124) that fluctuation effects are more important in materials with a small coherence length ξ. It is believed that the qualitative difference between the phase transitions to He II and to superconductivity are a consequence of the fact that the coherence length is only of order 1 nm in He II, and much longer in a superconductor. A numerical estimate for the temperature interval ΔT in which fluctuations are important in He II (Problem 6.20) gives $\Delta T \sim 0.5$K, in good agreement with the breadth of the λ anomaly (Fig. 1.4). For a pure superconductor, on the other hand, ΔT is very small, which is consistent with the fact that the transition to superconductivity gives an 'ideal' specific heat discontinuity (Fig. 1.12). We note also that although fluctuation effects may be expected to be small in any case in superconductors, the inequality (6.124) predicts that they are more noticeable in alloys, since ξ decreases as the mean free path l decreases.

The analogy with a magnetic system, which we drew in §6.9, points up an interesting feature of the phase change to superconductivity. Well above the Curie temperature T_c of a magnet, the magnetic susceptibility χ is given by the Curie–Weiss Law

$$\chi = \chi_0 T_c/(T - T_c), \tag{6.125}$$

where χ_0 is a constant. The susceptibility becomes very large as T approaches T_c.

The reason is that as T approaches T_c the spin system gets nearer to spontaneous alignment, so the spins align more and more readily with an applied field. The divergence in χ signals the onset of the phase change.

The same sort of thing happens with superconductors; here the d.c. conductivity diverges above T_c. A calculation of the effect has to be based on the microscopic theory, and it is not yet completely clear how the calculation should be carried out. The form of the divergence depends on the dimensionality of the system; it is

$$\sigma - \sigma_N = A\left(\frac{T_c}{T - T_c}\right)^n, \tag{6.126}$$

with an exponent n equal to $\frac{1}{2}$ for bulk material, 1 for a film and $\frac{3}{2}$ for a wire. σ_N is the normal state conductivity. The meaning of 'film', for example, is as follows. One can extend the notion of coherence length $\xi(T)$ to temperatures above T_c, and $\xi(T)$ is symmetrical about T_c:

$$\xi(T) \propto |T - T_c|^{-1/2}. \tag{6.127}$$

'Film' then means that one dimension is small compared with $\xi(T)$.

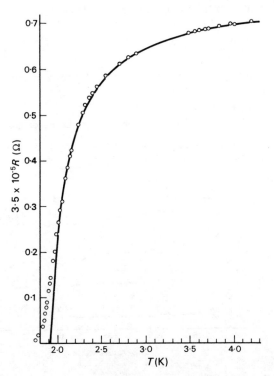

FIG. 6.38   Resistive transition of a granular Al film with a 'mean free path' l_{eff} of 0·2 Å, i.e. a normal state resistivity corresponding to this value. The theoretical curve is given by eq. (6.126), with $n = 1$ and T_c treated as a parameter. The A-value is 0·77 A_{th}, where A_{th} is given by eq. (6.128) (After Strongin et al., 1968.)

Most of the experimental work has been done on films. The original theory of Aslamazov and Larkin (1968) then gives for the constant A

$$A = e^2/16hd, \tag{6.128}$$

where d is the thickness of the film. In order to observe the 'paraconductivity' of eq. (6.126) it is necessary to work with very impure films. Although A in eq. (6.128) is independent of the mean free path l, the observability depends on $(\sigma - \sigma_N)/\sigma_N$, which increases as l decreases (Glover, 1971). Figure 6.38 shows the resistive transition in a granular Al film, one evaporated through an oxygen atmosphere so as to produce oxide barriers between the grains. Experimental results of this kind generally show the temperature dependence of eq. (6.126), and sometimes agree in magnitude with eq. (6.128). However, they are open to criticism (Testardi, 1971) on the grounds that the critical temperature T_c varies from grain to grain in such a film.

The conductivity is not the only property to show a singularity as T approaches T_c. Because a type I superconductor exhibits perfect diamagnetism, there is a magnetization singularity. The effect is very weak, but it has been measured by Gollub et al., (1969, 1970) with a SQUID on pure bulk indium specimens. An example of the change in magnetization above T_c is shown in Fig. 6.39.

In liquid helium, it is to be expected that fluctuations will lead to divergences in various transport coefficients near T_λ. Ferrell et al. (1968) have treated the

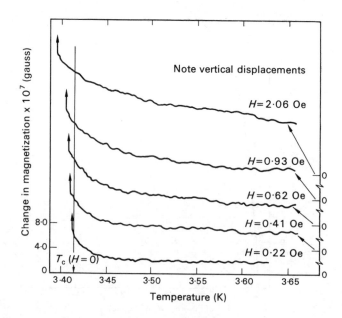

FIG. 6.39 Change in magnetization of a single crystal of indium (4·4 mm diameter) near the critical temperature, for various magnetic fields. Each curve terminates on the left end at the field-dependent thermodynamic critical temperature, where there is a sharp first-order transition to the superconducting state. (After Gollub et al., 1969.)

problem phenomenologically from the viewpoint of the scaling laws, which exploit the essential similarities between second-order phase transitions occurring in systems of widely differing character, for example ferromagnets, superconductors and liquid helium. In particular, they find that the thermal conductivity of He I should diverge as $(T - T_\lambda)^{-1/3}$. Measurements of the conductivity by Kerrisk and Keller (1969) support this prediction; Fig. 6.40 shows the variation of thermal conductivity with liquid density for various isotherms. Those which cross the λ-line on the p–T phase diagram are seen to rise sharply as the transition point is approached from above, whilst the isotherms that do not meet the λ-line exhibit only a slow variation.

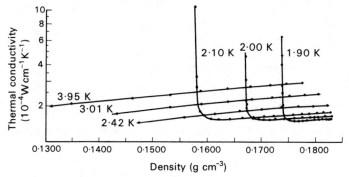

FIG. 6.40 Thermal conductivity of He I as a function of density for several temperatures. The three isotherms below $2.17\,\text{K}$ rise rapidly as the λ-line is approached. (Kerrisk and Keller, 1969.)

Other properties such as the damping of second sound (Tyson, 1968) also show a divergence at T_λ. Although some progress has been made in understanding the λ-transition, it is clear that more information, both experimental and theoretical, is needed.

Problems

6.1 Confirm that eq. (6.18) does follow from eq. (6.17). The following steps may be followed:

A: Expand the term

$$\frac{1}{2m} |(-i\hbar\nabla - 2e\mathbf{A})\psi|^2$$

in $g(\mathbf{r})$, and show that

$$\frac{\partial g}{\partial(\nabla_j\psi^*)} = \frac{i\hbar}{2m}(-i\hbar\nabla_j - 2eA_j)\psi$$

and

$$\frac{\partial g}{\partial\psi^*} = \alpha\psi + \beta|\psi|^2\psi - \frac{e}{m}\sum_j(-i\hbar\nabla_j - 2eA_j)\psi A_j.$$

B:　The condition div $\mathbf{A} = 0$ means that

$$\sum_j \nabla_j \psi A_j = \sum_j A_j \nabla_j \psi,$$

so that A_j can be brought to the front in the last term of $\partial g/\partial \psi^*$. Substitution of the above explicit expressions for $\partial g/\partial(\nabla_j \psi^*)$ and $\partial g/\partial \psi^*$ into the Euler–Lagrange equations then gives eq. (6.18) directly.

6.2　Confirm that the boundary condition of eq. (6.21) follows from eq. (6.20).

6.3　Confirm that eq. (6.23) follows from eq. (6.22). The following steps give one way:

A:　Expand the gradient term, and show that

$$\frac{\partial g}{\partial A_i} = -\frac{ie\hbar}{m}(\psi \nabla_i \psi^* - \psi^* \nabla_i \psi) - \frac{4e^2}{m} \psi^* \psi A_i.$$

B:　The x-component of the second term is

$$T_x = \frac{\partial}{\partial y}\frac{\partial g}{\partial(\partial A_x/\partial y)} + \frac{\partial}{\partial z}\frac{\partial g}{\partial(\partial A_x/\partial z)}.$$

There is no term in $\partial A_x/\partial x$, because this does not come into curl $\mathbf{A}$. Write out $(\text{curl } \mathbf{A})^2$ explicitly, and show that

$$\frac{\partial}{\partial y}\frac{\partial g}{\partial(\partial A_x/\partial y)} = \frac{1}{\mu_0}\left(\frac{\partial^2 A_x}{\partial y^2} - \frac{\partial^2 A_y}{\partial x \, \partial y}\right)$$

with a similar expression for the other term. Hence show that

$$T_x = \frac{1}{\mu_0}\left(\nabla^2 A_x - \frac{\partial}{\partial x}\text{div } \mathbf{A}\right)$$

$\left(\text{add and subtract } \dfrac{1}{\mu_0}\dfrac{\partial^2 A_x}{\partial x^2}\right)$. That is,

$$\mathbf{T} = \frac{1}{\mu_0}(\nabla^2 \mathbf{A} - \text{grad div } \mathbf{A}) = -\frac{1}{\mu_0}\text{curl curl } \mathbf{A}.$$

Equation (6.23) then follows. Note that this proof holds in a general gauge.

6.4　Use eqs. (6.30) and (6.31) and the defining equation for the entropy, $\Sigma = -(\partial G/\partial T)_H$ to show that the entropies in the normal and superconducting phases are respectively

$$\Sigma_n = -V\frac{\partial f_n}{\partial T}$$

$$\Sigma_s = \Sigma_n + V\frac{|\alpha|}{\beta}\frac{\partial|\alpha|}{\partial T},$$

and confirm that Σ_s is lower than Σ_n. Show that Σ_s can also be written

$$\Sigma_s = \Sigma_n + V\mu_0 H_{cb}\frac{\partial H_{cb}}{\partial T}.$$

Using the Gorter–Casimir temperature dependence $H_{cb} = H_0(1 - t^2)$, sketch $\Delta\Sigma = \Sigma_n - \Sigma_s$ as a function of temperature. Note that $\Delta\Sigma \to 0$ at $T = 0$, as required by the third law of thermodynamics.

A tin specimen passes adiabatically through the phase change at H_{cb} at a reduced temperature $t = 0.9$. What is the temperature drop? Use the following data for Sn: $T_c = 3.7\,\mathrm{K}$; $\mu_0 H_0 = 3 \times 10^{-2}\,\mathrm{T}$; specific heat $= 364\,\mathrm{J\,m^{-3}\,K^{-1}}$ at $t = 0.9$.

6.5 Kittel (quoted by de Gennes (1966)) pointed out that one could get a good approximation to the surface critical field H_{c3} by the variational method. The variational function $\psi(x) = \exp(-rx^2)$ satisfies the boundary condition of eq. (6.50), and is bounded as $x \to \infty$. Show that the variational integral

$$E = \int_0^\infty \left\{ \frac{\hbar^2}{2m}\left(\frac{d\psi}{dx}\right)^2 + \frac{2e^2\mu_0^2 H_0^2}{m}(x - x_0)^2\psi^2 \right\} dx \Big/ \int_0^\infty \psi^2\, dx$$

has the value

$$E = \frac{\hbar^2}{2m}r + \frac{2e^2\mu_0^2 H_0^2}{m}\left\{ \frac{1}{4r} - x_0\left(\frac{2}{\pi r}\right)^{1/2} + x_0^2 \right\}.$$

Minimize this expression with respect to r and x_0, to obtain

$$r = \frac{\mu_0 e H_0}{\hbar}\left(1 - \frac{2}{\pi}\right)^{1/2}$$

$$x_0 = \left(\frac{1}{2\pi r}\right)^{1/2}$$

$$E = \frac{\mu_0 e\hbar H_0}{m}\left(1 - \frac{2}{\pi}\right)^{1/2}.$$

The values of x_0 and E may be compared with the exact expressions of eqs. (6.52) and (6.53).

6.6 Confirm that the expressions for $H_{c\|}$ in the thick and thin limit, eqs. (6.54) and (6.58), do reduce to the forms quoted in eqs. (6.61) and (6.62).

6.7 Confirm the statement made in the text, that $|\psi|^2$ with ψ given by eq. (6.63) has zeros along the line $x = 0$ separated in y by λ_y. Show further that $|\psi|^2$ has no zeros except for $x = 0$.

6.8 Draw the line $\varepsilon = h$, corresponding to the critical field H_{c2}, on Fig. 6.14. Guyon et al. (1967) comment that the experimental curve for H_{fe} does not show any anomaly where it crosses the line $\varepsilon = h$. Why might one expect some anomaly?

6.9 Show that with the Abrikosov choice of all C_n's equal, ψ_L of eq. (6.71) satisfies the periodicity condition in x

$$\psi_L(x + k\xi^2(T), y) = \exp(iky)\psi_L(x, y)$$

and hence that $|\psi_L|^2$ is periodic in x with period $k\xi^2(T)$, as well as being periodic in y with period $2\pi/k$. Abrikosov's square lattice obviously corresponds to taking both periods equal, so that

$$k = (2\pi)^{1/2}/\xi(T).$$

Show that with this choice of k, $|\psi_L|^2$ vanishes at the centre of the unit cell,

$$x = y = \tfrac{1}{2}(2\pi)^{1/2}\xi(T)$$

and hence that $|\psi_L|^2$ vanishes on all the points of a square lattice. A set of level surfaces for this particular $|\psi_L|^2$ is shown in Fig. 6.19.

6.10 Confirm by direct substitution that eqs. (6.23) and (6.74) with (6.75) for the current $\mathbf{J}_L$ are equivalent for the wave function ψ_L with any values of C_n and k. (An alternative proof, using the factorization of the oscillator Hamiltonian into ladder operators, is given by de Gennes (1966), for example).

6.11 Equation (6.81) is a condition giving the normalization of ψ_L. Show that it can also be obtained as follows (de Gennes, 1966). In the expression for the free energy in terms of ψ_L, eqs. (6.15) and (6.16), write $\psi = (1+\varepsilon)\psi_L$. Show that eq. (6.81) is the condition for the coefficient of ε in the free energy to vanish.

6.12 Show that the first term in eq. (6.81) may be written

$$\int \frac{1}{2m}(i\hbar\nabla - 2e\mathbf{A}_{c2} - 2e\mathbf{A}_1)\psi_L^*(-i\hbar\nabla - 2e\mathbf{A}_{c2} - 2e\mathbf{A}_1)\psi_L \, d^3\mathbf{r}$$

and hence derive eq. (6.82).

6.13 Confirm that eq. (6.85) follows from eq. (6.82).

6.14 Use the identity

$$\left(\frac{\partial G}{\partial H}\right)_T = -M$$

and eq. (6.90) for M to show that

$$G(H_0) = G(H_{c2}) - \frac{\mu_0(H_{c2} - H_0)^2}{2\beta_A(2\kappa^2 - 1)}.$$

6.15 Show that eq. (6.102) follows from eq. (6.101). Confirm that $G_s = G_n$ for $f = 1$ and $H_0 = H_{cb}$ in the limit of large d, i.e. that the bulk limit is correct.

6.16 Show that the leading term in the expansion of G, eq. (6.102), is

$$G = f^2 \left\{ \frac{1}{3}\left(\frac{d}{2\lambda}\right)^2 \left(\frac{H_0}{H_{cb}}\right)^2 - 2 \right\},$$

and hence derive the critical supercooling field of eq. (6.107). Confirm by appropriate substitutions that this is the same as eq. (6.58).

6.17 Confirm eq. (6.108).

6.18 Confirm that eq. (6.118) follows from eq. (6.117).

6.19 Find an expression for the depression of the λ-point in a thin He II film of given thickness, using the GP theory. The depression predicted by the theory is rather low compared with experimental results.

6.20 Use eq. (6.124) to estimate the width ΔT of the fluctuation dominated region for (a) He II, (b) pure Sn.

References

ABRIKOSOV, A. A., 1957, *Zh. Eksper. Teor. Fiz.* **32**, 1442, (*Soviet Physics JETP* **5**, 1174).
ABRIKOSOV, A. A., GORKOV, L. P. and DZYALOSHINSKII, I. E., 1963, *Methods of Quantum Field Theory in Statistical Physics* (Pergamon, London).
ARP, V. D., COLLIER, R. S., KAMPER, R. A. and MEISSNER, H., 1966, *Phys. Rev.* **145**, 231.
ASLAMAZOV, L. G. and LARKIN, A. I., 1968, *Fiz. Tverdoga Tela* **10**, 1104. (*Soviet Physics Solid State* **10**, 875.)
BALDWIN, J. P., 1963, *Phys. Lett.* **3**, 223.
BALDWIN, J. P., 1964, *Rev. Mod. Phys.* **36**, 317.
CAROLI, C., CYROT, M. and DE GENNES, P. G., 1966, *Solid State Comm.* **4**, 17.
COLLIER, R. S. and KAMPER, R. A., 1966, *Phys. Rev.* **143**, 323.
DE GENNES, P. G., 1964, *Phys. Kond. Materie.* **3**, 79.
DE GENNES, P. G., 1966, *Superconductivity of Metals and Alloys* (Benjamin, New York).
DOLL, R. and GRAF, P., 1967, *Phys. Rev. Lett.* **19**, 897.
DOUGLASS, D. H., Jr., 1961, *Phys. Rev. Lett.* **7**, 14.
EILENBERGER, G., 1967, *Phys. Rev.* **153**, 584.
ESSMANN, U. and TRÄUBLE, H., 1967, *Phys. Lett.* **24A**, 526.
FABER, T. E., 1952, *Proc. Roy. Soc.* **A214**, 392.
FABER, T. E., 1955, *Proc. Roy. Soc.* **A231**, 353.
FABER, T. E., 1957, *Proc. Roy. Soc.* **A241**, 531.
FEDER, J. and MCLACHLAN, D. S., 1969, *Phys. Rev.* **177**, 763.
FERRELL, R. A., MENYHÁRD, N., SCHMIDT, H., SCHWABL, F. and SZÉPFALUSY, P., 1968, *Ann. Phys.* (N.Y.) **47**, 565.
FIETZ, W. A. and WEBB, W. W., 1967, *Phys. Rev.* **161**, 423.
GINZBURG, V. L., 1958, *Zh. Eksper. Teor. Fiz.* **34**, 113. (*Soviet Physics JETP* **7**, 78).
GINZBURG, V. L., 1960, *Fiz. Tverdoga Tela* **2**, 2031 (*Soviet Physics Solid State* **2**, 1824).
GINZBURG, V. L. and LANDAU, L. D., 1950, *Zh. Eksper. Teor. Fiz.* **20**, 1064.
GINZBURG, V. L. and PITAEVSKII, L. P., 1958, *Zh. Eksper. Teor. Fiz.* **34**, 1240. (*Soviet Physics JETP* **7**, 858).
GLOVER, R. E., 1971, *Superconductivity*, ed. F. S. Chilton (North-Holland), p. 3.
GOLLUB, J. P., BEASLEY, M. R., NEWBOWER, R. S. and TINKHAM, M., 1969, *Phys. Rev. Lett.* **22**, 1288.
GOLLUB, J. P., BEASLEY, M. R., and TINKHAM, M., 1970, *Phys. Rev. Lett.* **25**, 1646.
GORKOV, L. P., 1958, *Zh. Eksper. Teor. Fiz.* **34**, 735. (*Soviet Physics JETP* **7**, 505).
GORKOV, L. P., 1959a, *Zh. Eksper. Teor. Fiz.* **36**, 1918. (*Soviet Physics JETP* **9**, 1364).
GORKOV, L. P., 1959b, *Zh. Eksper. Teor. Fiz.* **37**, 1407. (*Soviet Physics JETP* **10**, 998).
GUYON, E., MEUNIER, F. and THOMPSON, R. S., 1967, *Phys. Rev.* **156**, 452.
GYGAX, S. and KROPSCHOT, R. H., 1964, *Phys. Lett.* **9**, 91.
HURAULT, J. P., 1966, *Phys. Lett.* **20**, 587.
JAIN, M. C. and MACKINNON, L., 1970, *Phys. Lett.* **32A**, 275.
KERRISK, J. F. and KELLER, W. E., 1969, *Phys. Rev.* **177**, 341.
KHALATNIKOV, I. M., 1969, *Zh. Eksper. Teor. Fiz.* **57**, 489. (*Soviet Physics JETP* **30**, 268).
KLEINER, W. H., ROTH, L. M. and AUTLER, S. H., 1964, *Phys. Rev.* **133A**, 1226.
MCCONVILLE, T. and SERIN, B., 1965, *Phys. Rev.* **140A**, 1169.
MAKI, K., 1964, *Physics* **1**, 21.
MAMALADZE, YU. G. and CHEISHVILI, O. D., 1966, *Zh. Eksper. Teor. Fiz.* **50**, 169. (*Soviet Physics JETP* **23**, 112).
PARKS, R. D., 1969, *Superconductivity*, 2 vols. (Dekker, New York).
PITAEVSKII, L. P., 1958, *Zh. Eksper. Teor. Fiz.* **35**, 408. (*Soviet Physics JETP* **8**, 282).
SAINT-JAMES, D. and DE GENNES, P. G., 1963, *Phys. Lett.* **7**, 306.
SHAPIRA, Y. and NEURINGER, L. J., 1965, *Phys. Rev.* **140**, A1638.
STRONGIN, M., KAMMERER, O. F., CROW, J., THOMPSON, R. S. and FINE, H. L., 1968, *Phys. Rev. Lett.* **20**, 922.

SUTTON, J., 1966, *Proc. Phys. Soc.* **87,** 791.
TESTARDI, L. R., 1971, *Phys. Lett.* **35A,** 33.
TILLEY, D. R., BALDWIN, J. P. and ROBINSON, G., 1966, *Proc. Phys. Soc.* **89,** 645.
TYSON, J. A., 1968, *Phys. Rev. Lett.* **21,** 1235.

Chapter 7

Pairing Effects in Other Systems

7.1 Introduction

We have now completed our discussion of He II and superconductors. In this final chapter we turn to three very different systems in which BCS type states may be found.

We pointed out in § 1.1 that He^4 is the only boson liquid which does not solidify under atmospheric pressure as its temperature is lowered. For this reason boson condensation is unique to He^4. On the other hand, there are systems of interacting fermions in which condensation of the BCS type is possible; we recall from § 1.2 that two criteria are that the system should be degenerate, that is have a Fermi surface, and that there should be an attractive interaction between particles at the Fermi surface. These criteria are satisfied in nuclei and in some parts of neutron stars, and may be satisfied in liquid He^3.

7.2 Liquid He^3

As we stressed in Chapter 1, liquid He^3 is a fermion system, and does not undergo a λ-transition like liquid He^4. However, for a number of years there has been speculation that He^3 might undergo a BCS type of condensation, and various attempts have been made, for example by Emery and Sessler (1960), to predict a transition temperature.

The BCS state we describe in Appendix 1 is composed of Cooper pairs in which the electrons have opposed spin. Because the total wave function of the system must change sign when the coordinates of two electrons are exchanged, this means that the relative spatial wave function of the electrons in a pair must be symmetric. The simplest possibility is that the spatial wave function is isotropic, that is an S-wave, and it is believed that superconductivity in metals always involves S-wave pairing. In He^3, however, the interparticle force is anisotropic and in addition it is repulsive at a short enough distance. It is therefore generally assumed that if a BCS type of state is to occur in liquid He^3, it will involve pairs in something other than an S-wave, for example a P-wave with parallel spins.

The results of experiments by Osheroff et $al.$ (1972a, b) on He^3 along the melting curve have been interpreted as indicating that a BCS state may indeed be formed. Using an adiabatic compressional cooling apparatus, these authors found two anomalies in the pressure–time curve, (Fig. 7.1), when the volume was changed at a constant rate. At 2·7 mK the slope of the curve suffered a discontinuity (point A),

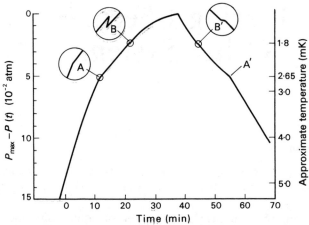

FIG. 7.1 Time evolution of melting pressure of He^3 during compression and subsequent decompression. A, A'—anomalies showing discontinuity in slope; B—anomaly with hysteresis; B'—anomaly occurs at slightly lower pressure than B, no hysteresis. Estimated temperatures plotted at right (Osheroff et $al.$, 1972b.)

and at about 1·8 mK there was a singularity involving hysteresis (point B). At first these features were thought to indicate phase changes in solid He^3. However, it was found that the nuclear magnetic resonance signal from the liquid phase is equal to that from the solid above point A, but below A the liquid signal increases to higher values than the solid signal, until, at point B, it suddenly returns to its original value. Leggett (1972) has suggested that point A marks a second-order phase transition of the liquid into a BCS state, in which the He^3 spins are paired to form a triplet state and the pairs have odd values of relative angular momentum. Using a sum-rule argument, Leggett finds qualitative agreement with the thermodynamic and frequency shift data. Further support for the occurrence of a second-order phase change in the liquid has come from specific heat measurements by Webb et $al.$ (1973). The specific heat of liquid He^3 shows a discontinuity, but not a divergence, in the temperature range 2 to 3 mK, at pressures varying from 16 atm to the melting pressure (34 atm). It seems probable that this discontinuity is associated with anomaly A (Fig. 7.1).

Ambegaokar and Mermin (1973) have adopted another theoretical approach to the problem; starting with a Ginzburg–Landau type free-energy functional for He^3 pairing in states of odd angular momentum, they make several predictions about the H–T phase diagram. In particular, it is suggested that anomaly A should suffer linear splitting in a magnetic field.

Anderson and Varma (1973) have attempted to give a microscopic picture of the liquid phase between points A and B, and have found an expression for the n.m.r. frequency shift which has the correct temperature dependence. In a further paper Anderson (1973) emphasizes the complex nature of the proposed anisotropic superfluid state. He notes that there are three vectors to be considered in a pair of He^3 atoms, the nuclear magnetic moment, the total spin and the relative angular momentum. The nature of the liquid phases will be determined by the relationships amongst these vectors. In particular, Anderson suggests that between points A and B, the spin of a pair can take only two values, even though it is in a triplet state, whilst point B marks a first-order transition to the state in which all three spin settings are allowed.

7.3 Finite nuclei

The protons and neutrons, collectively called nucleons, in the atomic nucleus may be regarded as filling a set of shells, just like the electrons in the atom. (For a discussion of the nuclear shell model see Elton, 1965.) The shell description is not perfect, and there are residual interactions between the nucleons which among other things lead to the formation of Cooper pairs of nucleons. The nucleons in a pair have opposite angular momenta, so that angular momentum takes over the role played by linear momentum in a pure superconductor. Because the nucleus contains only a small and fixed number of particles, pairing leads to a systematic increase δB in the binding energy B of nuclei with an even number N of neutrons or an even number Z of protons. It is found that

$$\delta B = \begin{cases} \Delta & N \text{ even } Z \text{ even} \\ 0 & N+Z \text{ odd} \\ -\Delta & N \text{ odd } Z \text{ odd,} \end{cases} \tag{7.1}$$

where δB is the deviation from the average binding energy $B(N, Z)$ of nuclei with N- and Z-values near those of the given nucleus. To be precise, the neutron gap Δ_n can be estimated for even N as

$$\Delta_n = \tfrac{1}{4}\{B(N-2, Z) - 3B(N-1, Z) + 3B(N, Z) - B(N+1, Z)\}. \tag{7.2}$$

The proton gap Δ_p is estimated in a similar way. Fig. 7.2 shows the marked effect of pairing. The values of Δ_n and Δ_p may be compared with the total binding energy per nucleon, which is of order 8 MeV.

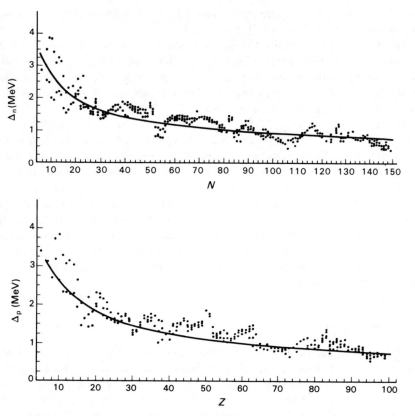

FIG. 7.2   (a) The neutron gap Δ_n and (b) the proton gap Δ_p. The empirical curve $\Delta = 12A^{-1/2}$ is drawn in each case, where $A = N + Z$ is the total nucleon number. (After Bohr and Mottelson, 1969.)

A second consequence of pairing is that the ground states of even–even nuclei (even N and even Z) have zero angular momentum, since the nucleons are all paired up. Thirdly, to create an excitation in an even–even nucleus involves breaking a pair, so the lowest excited states lie systematically higher in even–even nuclei than in other nuclei.

7.4   Neutron stars

The possibility that a neutron star might be the end product of a dying star, whose mass is similar to that of the sun, was first suggested by Gamow (1936). When all the thermonuclear fuel in the star has been used up, gravitational collapse commences and continues until, under certain conditions, equilibrium is established between the gravitational forces and the internal pressure of the 'neutron liquid' which is formed inside the star. The probable structure of such a neutron star is shown in Fig. 7.3. The bulk of the star consists of neutrons at the

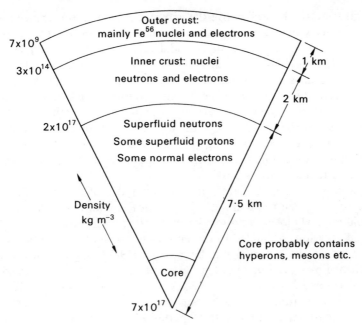

Outer crust:
mainly Fe^{56} nuclei and electrons

7×10^{9}

3×10^{14}

Inner crust: nuclei

neutrons and electrons

1 km

2 km

2×10^{17}

Superfluid neutrons

Some superfluid protons

Some normal electrons

Density
kg m^{-3}

7·5 km

Core probably contains
hyperons, mesons etc.

Core

7×10^{17}

FIG. 7.3   Cross-section of neutron star. (After Pines, 1971.)

huge density of $\sim 10^{17}$ kg m^{-3}; its mass is expected to be still comparable with that of the sun ($\sim 10^{30}$ kg), so that its radius is 10 km, very small compared with the size of the original star.

Little interest was shown in the possible existence of neutron stars until Gold (1968) proposed that the recently discovered pulsars might be rotating neutron stars. The chief characteristic of a pulsar is that it emits short pulses of radiation at very regular intervals; the presently known pulsars have periods ranging from 33 ms to a few seconds. The pulsar period is identified with the rotational period of the neutron star, but a satisfactory explanation of the mechanism which gives rise to the radiation has still to be worked out. Almost certainly, it originates from the motion of charged particles in the intense magnetic field ($\sim 10^{8}$ T) surrounding the star, but this is a complex problem and not our concern here.

Neutron stars have recently received attention from low-temperature physicists because of the likelihood that the neutrons (and the protons) form some kind of superfluid phase. Neutrons at a density of $\sim 10^{17}$ kg m^{-3} comprise a fermion system with a degeneracy temperature $\sim 10^{11}$ K. On the other hand, the temperature of the star is probably $\sim 10^{8}$ K, so that the neutrons are highly degenerate. Migdal (1959) was the first to propose that the neutrons near the Fermi surface might be paired in such a way as to suffer a BCS-type condensation. According to Hoffberg et al. (1970), there is a critical neutron density ($1\cdot45 \times 10^{17}$ kg m^{-3}) below which S-wave pairing is dominant, as in superconductors, but above which P-wave pairing takes over, as has been suggested for He^{3}. Using the standard BCS relations, these authors estimate that the transition temperature for the neutron superfluid is $\sim 10^{10}$ K, well above the actual star temperature. Since the major part of the mass of the star consists of neutrons, it appears that most of the rotational energy resides in the superfluid. By analogy with rotating He II, Ginzburg and Kirzhnits (1964) concluded that the neutron superfluid would contain an array of vortex lines. In the same way as in He II, one can define critical values of angular velocity, Ω_{c1}, which must be exceeded to form a single vortex, and Ω_{c2}, at which the vortex cores would overlap. Baym et al. (1969) give the values $\Omega_{c1} \sim 10^{-14}$ s^{-1} and $\Omega_{c2} \sim 10^{20}$ s^{-1}; the periods of all the known pulsars correspond to rotational speeds (Ω) ranging from about 1 s^{-1} to 191 s^{-1}. We conclude that the neutron superfluid has properties similar to He II undergoing solid-body rotation at a temperature ~ 1 mK.

The pulsar periods are found to increase slowly at a constant rate. This suggests that the rotating neutron stars are slowing down, and, if so, that the neutron superfluid is slowly losing rotational energy. The problem then arises of how the superfluid rotation is coupled to the process in the surrounding magnetic field which results in the pulsar signal. Baym et al. (1969) suggest that this must be accomplished by scattering of the protons and electrons off the vortex cores. The fractional volume occupied by the vortex cores is approximately $\Omega/\Omega_{c2} \sim 10^{-18}$ for the fastest rotation, and as a result the coupling time for communicating changes in rotational speed from the superfluid to the crust is of the order of years, which is the kind of magnitude required to explain the observations.

A comprehensive review of neutron star properties can be found in the article by Pines (1971).

References

AMBEGAOKAR, V. and MERMIN, N. D., 1973, Phys. Rev. Lett. **30**, 81.

ANDERSON, P. W., 1973, Phys. Rev. Lett. **30**, 368.

ANDERSON, P. W. and VARMA, C. M., 1973, Nature (Lond.) **241**, 187.

BAYM, G., PETHICK, C. J. and PINES, D., 1969, Nature (Lond.) **224**, 673.

BOHR, A. and MOTTELSON, B. R., 1969, Nuclear Structure (W. A. Benjamin, New York).

ELTON, L. R. B., 1965, Introductory Nuclear Theory (Pitman, London), 2nd. edn.

EMERY, V. J. and SESSLER, A. M., 1960, Phys. Rev. **119**, 43.

GAMOW, G., 1936, Atomic Nuclei and Nuclear Transformations (Oxford University Press, London), 2nd. edn. p. 234.

GINZBURG, V. L. and KIRZHNITS, D. A., 1964, Zh. Eksper. Teor. Fiz. **47**, 2006. (Soviet Physics JETP **20**, 1346).

GOLD, T., 1968, Nature (Lond.) **218**, 731.

HOFFBERG, M., GLASSGOLD, A. E., RICHARDSON, R. W. and RUDERMAN, M., 1970, Phys. Rev. Lett. **24**, 775.

LEGGETT, A. J., 1972, Phys. Rev. Lett. **29**, 1227.

MIGDAL, A. B., 1959, Nucl. Phys. **13**, 655.

OSHEROFF, D. D., GULLY, W. J., RICHARDSON, R. C. and LEE, D. M., 1972a, Phys. Rev. Lett. **29**, 920.

OSHEROFF, D. D., RICHARDSON, R. C. and LEE, D. M., 1972b, Phys. Rev. Lett. **28**, 885.

PINES, D., 1971, Proc. 12th Int. Conf. on Low Temp. Phys., Kyoto 1970 (Academic Press of Japan, Tokyo), p. 7.

WEBB, R. A., GREYTAK, T. J., JOHNSON, R. T. and WHEATLEY, J. C., 1973, Phys. Rev. Lett. **30**, 210.

Appendix 1
Superconductivity:
Description
of Microscopic Theory

As we mentioned in §1.2, the microscopic theory is based on the work of Bardeen *et al.* (1957), and this appendix is largely devoted to describing results from that paper.

The first great difficulty in the development of the theory was to discover the nature of the interaction responsible for the transition to superconductivity. It is now believed that the mechanism is a coupling between electrons via the positive ions of the metallic lattice. This mechanism was first proposed by Fröhlich (1950), and experimental evidence in support soon followed in the discovery of the *isotope effect*. It was found in the original experiments that the critical temperature of a mercury specimen depended on the average mass number M of the ions in the metal; since there are several stable isotopes of mercury it was possible to prepare a range of specimens with a range of M-values. This isotope effect is found in most superconductors, and it obeys the empirical law

$$T_c \propto M^{-a} \tag{A1.1}$$

The exponent a is 0·5, or slightly less, for most non-transition metals, but can be small for transition metal superconductors. A table of values is given in Lynton (1969).

The isotope effect makes it clear that the ions in the metal play an essential role in superconductivity. Fröhlich suggested that the interaction between electrons and lattice vibrations, or phonons, could lead to an effective interaction between the electrons themselves. The electrons are coupled to the lattice vibrations because the ions of the lattice are positively charged; this coupling to the lattice means that an electron can emit a phonon, that is set the ions into vibration, and it can also absorb a phonon. Suppose one calculates the energy of a pair of electrons travelling through the lattice, using quantum-mechanical perturbation theory. A possible intermediate state is one in which a phonon is emitted by one of the electrons and absorbed by the other, as illustrated in Fig. A1.1. Inclusion of this

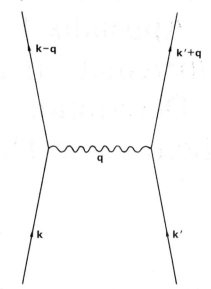

FIG. A1.1　　Phonon exchange between electrons.

intermediate state means that the energy of the two electrons is altered, so the phonon exchange has the same effect as a direct interaction between the electrons. The equivalent direct interaction for the exchange of one phonon turns out to be

$$V(\mathbf{k}, \mathbf{k}', \mathbf{q}) = \frac{g^2 \hbar \omega_q}{(\varepsilon_{k+q} - \varepsilon_k)^2 - (\hbar \omega_q)^2} \tag{A1.2}$$

where, following Fig. A1.1, the momenta of the incoming electrons are $\hbar\mathbf{k}$ and $\hbar\mathbf{k}'$, and the momentum of the exchanged phonon is $\hbar\mathbf{q}$. In eq. (A1.2) ε_k is the energy of the electron state with wave number $\mathbf{k}$, $\hbar\omega_q$ is the phonon energy, and g is the coupling constant for the interaction between the electrons and the phonons.

The important thing about eq. (A1.2) is that the interaction is negative, that is to say attractive, for $|\varepsilon_{k+q} - \varepsilon_k| < \hbar\omega_q$. We must also take into account the Coulomb (electrostatic) force between the electrons, which is always repulsive. Depending on their relative strengths, the sum of the two forces can be either attractive or repulsive; we shall see that superconductivity arises if the net force is attractive. In fact, the superconducting state has lower energy than the normal state, so it can only be produced by an attractive force. It is because the sign of the interaction depends on a delicate balance between the phonon-induced attraction and the Coulomb repulsion that there is no simple rule for deciding whether a given material becomes superconducting. Note, however, that the attractive interaction, eq. (A1.2), depends on g^2, where g is the electron–phonon coupling constant, so that it is favourable to superconductivity to have strong coupling between the electrons and phonons. The simplest consequence of strong electron–phonon coupling is that the room-temperature resistivity is high, since it is determined by electron–phonon scattering. This is why the noble metals are not superconductors: they are good conductors at room temperature, so that g is small, and Coulomb repulsion can be expected to dominate the phonon-exchange interaction.

Even though eq. (A1.2) is derived simply from one phonon exchange, it is still complicated, and becomes more complicated when the Coulomb interaction is added. The transition to superconductivity arises from the attractive part of the interaction, and the main features can be derived from the simplified interaction used by Bardeen *et al.*:

$$V_{kk'} = \begin{cases} -V & \text{if } |\varepsilon_k| < \hbar\omega_D \text{ and } |\varepsilon_{k'}| < \hbar\omega_D \\ 0 & \text{otherwise,} \end{cases} \qquad (A1.3)$$

where $\hbar\omega_D$ is the Debye energy for phonons in the lattice. $V_{kk'}$ is the interaction between electrons which before scattering have momenta $\hbar k$ and $\hbar k'$, and after scattering have momenta $\hbar k_1$ and $\hbar k_1'$, say (see Fig. A1.1); conservation of momentum requires

$$\hbar k_1 + \hbar k_1' = \hbar k + \hbar k' = \hbar K, \qquad (A1.4)$$

where $\hbar K$ is the momentum of the centre of mass of the electrons. The energies ε_k and ε_k are measured from the Fermi surface, so that ε_k is negative if k lies within the Fermi sphere, positive if ε_k lies outside. It is argued in setting up eq. (A1.3) that the interaction can only affect electrons in the neighbourhood of the Fermi surface, because electrons inside the Fermi sphere are prevented by the Pauli principle from scattering into any other states. Equation (A1.2) gives an attractive interaction provided $|\varepsilon_{k+q} - \varepsilon_k| < \hbar\omega_q$, so the simplified interaction is taken as attractive if both ε_k and $\varepsilon_{k'}$ are less than an average phonon energy, for which the Debye energy serves. In order of magnitude, comparison with eq. (A1.2) shows that the interaction constant V in eq. (A1.3) is about $g^2/\hbar\omega_D$. If the average mass number M is increased, the ions become less mobile, so both g and $\hbar\omega_D$ decrease. In fact the combination $g^2/\hbar\omega_D$ is independent of M, so in seeking an explanation for the isotope effect, V can be taken independent of M.

The interaction of eq. (A1.3), and the assumption that the Fermi surface is spherical, are the main ingredients of the BCS theory in its simplest form.

Equation (A1.3) describes an interaction between electrons in a shell within an energy $\hbar\omega_D$ of the Fermi surface. $\hbar\omega_D$ is always much less than the Fermi energy ε_F; typically $\hbar\omega_D/\varepsilon_F$ is about 10^{-3}. Hence the shell is a very thin one. The interaction leaves the centre-of-mass momentum $\hbar K$ of two interacting electrons unaltered, and one can see that the interaction is most effective if $\hbar K = 0$. Suppose we look for the effect of the interaction on pairs with a general centre-of-mass momentum $\hbar K$; for the interaction to involve a given pair they must lie within the shell of thickness $\hbar\omega_D$ in energy. To find the pairs of points in momentum space which lie within the interaction shell and have centre-of-mass momentum $\hbar\mathbf{K}$, one can use the following construction (Fig. A1.2). Draw the Fermi sphere, radius k_F,

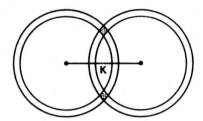

FIG. A1.2   Construction to find states within energy shell and with centre-of-mass momentum $\hbar\mathbf{K}$.

say, with a shell, then draw the same sphere displaced by $\mathbf{K}$. The required points lie in the region of intersection of the two shells.

The construction in Fig. A1.2 shows that the number of electron levels involved in the interaction increases as $\mathbf{K}$ decreases, and has a sharp maximum for $\mathbf{K} = 0$ when the whole shell is involved. The effects of the interaction are therefore strongest between electrons of opposite momentum, $\mathbf{k}$ and $-\mathbf{k}$.

In working out the consequences of the BCS interaction, it is vital to take account of the Pauli exclusion principle. Cooper (1956) solved the Schrödinger equation for a pair of electrons of opposite momenta $\mathbf{k}$ and $-\mathbf{k}$ in the 'potential well' of eq. (A1.3); he incorporated the exclusion principle approximately by specifying $|\mathbf{k}| > k_F$, that is, by assuming that the interacting electrons could not get into the Fermi sphere. The result is that however weak the interaction V is, the electron pair always has a bound state, with binding energy E_b given by

$$E_b = 2\hbar\omega_D \exp\{-2/N(0)V\}, \tag{A1.5}$$

in which $N(0)$ is the density of states in energy at the Fermi surface. That is, $N(0)\,\delta E$ is the number of electron levels (neglecting spin) with energy between E_F and $E_F + \delta E$. Cooper found for the radius of the bound state

$$R \sim \hbar^2 k_F / m E_b. \tag{A1.6}$$

Numerically R is of order 10^{-6} m, an enormous length on an atomic scale.

Equation (A1.5) for the binding energy of a Cooper pair is of the right magnitude for a theory of superconductivity. A Cooper pair should be a stable structure as long as the thermal energy of the electrons is small enough, that is, for $k_B T < E_b$. So if the superconducting state is built of Cooper pairs in some sense, we may expect $k_B T_c \sim E_b$. The ratio $k_B T_c/\hbar\omega_D$ is much less than one for all materials; this result is now easy to understand, as the exponential in eq. (A1.5) can be very small.

One further significant feature of Cooper's calculation is that in the bound state the wave function $\psi(\mathbf{k})$ of the pair depends only on $|\mathbf{k}|$, the magnitude of $\mathbf{k}$. This means that the wave function is symmetric if the positions of the electrons are exchanged, so automatically it is antisymmetric in the spins of the electrons. That is, the spins are in opposite directions.

The principal defect of Cooper's calculation is that it singles out two particular electrons and treats them quite differently from all the other electrons. This defect is remedied in the BCS theory, which treats all the electrons on the same footing. The BCS theory takes as Hamiltonian the kinetic energy of all the electrons, together with the interaction of eq. (A1.3). The ground-state energy is calculated by the variational method: the variational wave function is constructed so that if one member $(\mathbf{k}, \uparrow)$ of a Cooper pair is present, so is $(-\mathbf{k}, \downarrow)$, the other member (the arrows indicate the spin wave functions). The state so constructed is different from the normal state. Let $h_{\mathbf{k}}$ be the probability that the pair $(\mathbf{k}, \uparrow)$ and $(-\mathbf{k}, \downarrow)$ is occupied. In the normal state, $h_{\mathbf{k}}$ is 1 for $|\mathbf{k}|$ inside the Fermi sphere, and 0 for $|\mathbf{k}|$ outside. In the superconducting state $h_{\mathbf{k}}$ is rounded through the interaction shell—see Fig. A1.3. The kinetic energy is allowed to increase relative to the normal state, so as to take advantage of the attractive interaction. It is found that

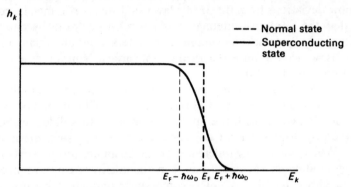

FIG. A1.3   Occupation probability for Cooper pairs in the BCS ground state.

the variational state does have lower energy than the normal state: the energy relative to the normal state at 0 K is

$$W(0) = -\frac{2N(0)(\hbar\omega_{\mathrm{D}})^2}{\exp\{2/N(0)V\} - 1}. \tag{A1.7}$$

For small values of $N(0)V$ we need only the exponential in the denominator, and $W(0)$ can be written in terms of the binding energy E_{b} of the Cooper pair, eq. (A1.5), as

$$W(0) \sim -N(0)\hbar\omega_{\mathrm{D}}E_{\mathrm{b}}. \tag{A1.8}$$

This means $W(0)$ is simply the number of Cooper pairs, $N(0)\hbar\omega_{\mathrm{D}}$, in the interaction shell, multiplied by the binding energy of each pair.

We can see how some of the other properties of the BCS state correlate with properties of superconductors. The first, obvious, comment is that in both the Cooper pair and the BCS ground state it is assumed that there is a Fermi surface: formally $N(0)$ in eqs. (A1.5) and (A1.7) must not be zero. This means that the calculation can only apply to metals, and we saw that only metals become superconducting. There is a class of materials known as superconducting semiconductors (for example $SrTiO_3$), but these are semiconductors which have been so heavily doped that the Fermi energy has risen into the conduction band, so that $N(0)$ is non-zero.

We argued already that the critical temperature is given by $k_B T_c \sim E_b$. By minimizing an appropriate free energy rather than the internal energy, one can generalize the BCS calculation to finite temperature. The critical temperature is that at which the free energy of the BCS state equals the normal state free energy; it is found to be

$$k_B T_c = 1 \cdot 14 \hbar \omega_D \exp \{ -1/N(0)V \}. \tag{A1.9}$$

Equation (A1.9) contains the isotope effect. Dimensionally, ω_D must be given as $\omega_D \sim (K/M)^{1/2}$, where K is a force constant of the lattice. We already remarked that V should be independent of M, so eq. (A1.9) does give $T_c \propto M^{-1/2}$. We can also see how deviations from the $M^{-1/2}$ law can arise for real metals: eq. (A1.2) assumes that the attractive interaction is given satisfactorily by one phonon exchange; eq. (A1.3) drastically simplifies the interaction; and finally the whole calculation is carried out for a spherical Fermi surface. None of these simplifications need hold for a real metal.

We remarked in § 1.2 that the addition of magnetic impurities rapidly destroys superconductivity. The reason for this is that in the neighbourhood of a magnetic impurity the interaction with the impurity spin leads to a difference in energy between the two electrons in a Cooper pair because their spins are in different directions. When this energy difference, or rather an appropriate average of it, exceeds the binding energy of the pair, superconductivity can no longer occur. Equally, magnetic metals do not become superconductors because the two electrons of a pair would have widely different energy. Finally, we can see how the *paramagnetic limit* on very high critical fields arises. The Pauli paramagnetism can be understood in terms of a splitting of the Fermi sphere between electrons with spin in the direction of the magnetic field and electrons with opposite spin; because of the interaction with the applied magnetic field the two sets have different energy. Again, superconductivity disappears when the difference in energy between up spin and down spin states at the Fermi surface exceeds the binding energy of the Cooper pair.

We saw that ordinary impurities, unlike magnetic impurities, do not alter the critical temperature much. The reason for this was first explained by Anderson (1959). In the normal state, the electrons may be described by wave functions $\phi_{n\uparrow}(\mathbf{r})$, $\phi_{n\downarrow}(\mathbf{r})$, where ϕ_n is supposed to include the effects of the impurity scattering. The quantum number n replaces the wave number $\mathbf{k}$ we used for the pure metal. The Cooper pair in the pure metal is composed of the states $(\mathbf{k}\uparrow)$ and $(-\mathbf{k}\downarrow)$; Anderson pointed out that the second of these states is the first with momentum and current reversed in time. He argued that in the impure metal one should equally pair time-reversed states, namely $\phi_{n\downarrow}(\mathbf{r})$ and $\phi_{n\downarrow}^*(\mathbf{r})$: the complex conjugate ϕ_n^* is the time reverse of ϕ_n just as $e^{-i\mathbf{k}\cdot\mathbf{r}}$ is the time reverse of $e^{i\mathbf{k}\cdot\mathbf{r}}$. With non-magnetic impurities, ϕ_n^* and ϕ_n have the same energy, and the BCS calculation goes through unmodified. In fact the impurity scattering washes out the effects of Fermi surface anisotropy, so that the BCS results apply rather better to alloys than to pure metals. We saw (Fig. 1.13) that when impurities are added to a pure superconductor, T_c first drops sharply, then varies only slowly with impurity content. The sharp drop is connected with the destruction of anisotropy effects and the subsequent slow change is due to the change of $N(0)$ because of 'doping' by the impurities.

To conclude this section, we mention one of the most important results of the BCS theory, the existence of an energy gap in the excitation spectrum of the superconductor. The ground state of a normal metal at 0 K is simply the state in which all the electron levels inside the Fermi sphere are occupied, and all the levels outside are unoccupied. A simple excited state is one in which an electron is removed from an occupied level $\mathbf{k}_1$ and placed in an unoccupied level $\mathbf{k}_2$; the energy required is

$$E(\mathbf{k}_1,\mathbf{k}_2) = |\varepsilon_{\mathbf{k}_1}| + |\varepsilon_{\mathbf{k}_2}|, \tag{A1.10}$$

where as before $\varepsilon_{\mathbf{k}}$ is the energy measured from the Fermi surface. By taking $\mathbf{k}_1$ and $\mathbf{k}_2$ close enough to the Fermi surface, $E(\mathbf{k}_1,\mathbf{k}_2)$ can be made arbitrarily small, so there is no gap in the excitation spectrum. For the superconductor, the BCS theory leads to the result that $|\varepsilon_{\mathbf{k}}|$ must be replaced by $E_{\mathbf{k}}$, with

$$E_{\mathbf{k}} = (\varepsilon_{\mathbf{k}}^2 + \Delta^2)^{1/2}. \tag{A1.11}$$

The lowest excitation energy is now 2Δ: there is a *gap* in the excitation spectrum. To create an excitation, it is necessary to break up a Cooper pair, so that the energy gap must be about equal to the binding energy of a pair, $\Delta \sim E_b$. The exact result at zero temperature is

$$2\Delta(0) = 4\hbar\omega_D \exp\{-1/N(0)V\} = 3\cdot52 k_B T_c, \tag{A1.12}$$

where the second part follows from eq. (A1.9).

The free-energy difference between the superconducting and normal states goes to zero at T_c; consequently E_b and Δ must go to zero at T_c, so that Δ is temperature dependent, $\Delta = \Delta(T)$. The variation of Δ as a function of T is shown in Fig. 1.25.

References

ANDERSON, P. W., 1959, *J. Phys. Chem. Solids* **11**, 26.
BARDEEN, J., COOPER, L. N. and SCHRIEFFER, J. R., 1957, *Phys. Rev.* **108**, 1175.
COOPER, L. N., 1956, *Phys. Rev.* **104**, 1189.
FRÖHLICH, H., 1950, *Phys. Rev.* **79**, 845.
LYNTON, E. A., 1969, *Superconductivity*, 3rd edition (Methuen, London).

Bibliography

I. Sources including both liquid helium and superconductivity

1. London, F. *Superfluids*, Vol. I (*Superconductivity*) and Vol. II (*Superfluid Helium*) (Wiley, New York, 1954; reprinted by Dover, New York, 1964).
   The classic work which first stressed the similarities between superconductivity and superfluidity.

2. *Progress in Low Temperature Physics* (ed. C. J. Gorter), Vol. I (1955), Vol. II (1957), Vol. III (1961), Vol. IV (1964), Vol. V (1967), Vol. VI (1970). (North-Holland, Amsterdam).
   Each volume contains a series of review articles on various aspects of low-temperature physics.

3. *Proceedings of the International Conferences on Low-Temperature Physics* (LT),
   LT 7 Toronto 1960, ed. G. M. Graham and A. C. Hollis Hallett (University of Toronto Press, 1961).
   LT 8 London 1962, ed. R. O. Davies (Butterworth, London, 1963).
   LT 9 Columbus 1964, ed. J. G. Daunt, D. O. Edwards, F. J. Milford, and M. Yaqub (Plenum Press, New York, 1965).
   LT 10 Moscow 1966, ed. M. P. Malkov (Viniti, Moscow, 1967).
   LT 11 St. Andrews 1968, ed. J. F. Allen, D. M. Finlayson and D. M. McCall (University of St. Andrews, 1968).
   LT 12 Kyoto 1970, ed. E. Kanda (Academic Press of Japan, Tokyo, 1971).
   LT 13 Boulder, 1972, ed. W. O'Sullivan, K. D. Timmerhaus and E. F. Hammel (Plenum Press, New York, 1973).

4. Brewer, D. F. (ed.), *Quantum Fluids* (Proceedings of the Sussex University Symposium 1965), (North-Holland, Amsterdam, 1966).
   Contains a number of valuable papers.

II. Liquid helium

1. Atkins, K. R., *Liquid Helium* (Cambridge University Press, London, 1959).
    A clear introduction to the subject, particularly the two-fluid model.
2. Careri, G. (ed.) *Liquid Helium, Proceedings of the Enrico Fermi International School of Physics* (course XXI), Varenna (Academic Press, London, 1963).
    A collection of review papers.
3. Khalatnikov, I. M., *Introduction to the Theory of Superfluidity*, translated by P. C. Hohenberg (W. A. Benjamin, New York, 1965).
    A concise account of the Russian contribution to the theory of liquid helium.
4. Wilks, J., *The Properties of Liquid and Solid Helium* (Clarendon Press, Oxford, 1967).
    A comprehensive survey, particularly of experimental work.
5. Donnelly, R. J., *Experimental Superfluidity* (University of Chicago Press, Chicago, 1967).
    In spite of its title, this book contains detailed discussion of various aspects of the theory, as well as an account of important experimental results.
6. Keller, W. E., *Helium-three and Helium-four* (Plenum Press, New York, 1969).
    A wide-ranging source-book.
7. Allen, J. F. (ed.), *Superfluid Helium* (Proceedings of the St. Andrews University Symposium 1965), (Academic Press, London, 1965).
    Contains papers on several aspects of superflow.

III. Superconductivity

A. *General books*

1. Shoenberg, D., *Superconductivity* (Cambridge University Press, London, 2nd edition, 1952).
    The standard pre-BCS account, with thermodynamics treated in detail, and a substantial treatment of the intermediate state.
2. Lynton, E. A., *Superconductivity* (Methuen, London, 3rd edition, 1969).
    Comprehensive and concise introductory account.
3. Schrieffer, J. R., *Theory of Superconductivity* (Benjamin, New York, 1964).
4. Rickayzen, G., *Theory of Superconductivity* (Wiley, New York, 1965).
    Both Schrieffer and Rickayzen contain a detailed treatment of the microscopic theory.
5. de Gennes, P. G., *Superconductivity of Metals and Alloys* (Benjamin, New York, 1966).
    A very stimulating book, primarily theoretical, with particularly useful sections on type II superconductivity and Ginzburg–Landau theory.

6. Rose-Innes, A. C. and Rhoderick, E. H., *Introduction to Superconductivity* (Pergamon, Oxford, 1969).
   A clear introductory work.
7. Kuper, C. G., *Theory of Superconductivity* (Clarendon, Oxford, 1969).
   A useful introduction to the theory.
8. Parks, R. D., *Superconductivity*, 2 Vols. (Dekker, New York, 1969).
   The standard detailed source work.
9. Williams, J. E. C., *Superconductivity and its Applications* (Pion, London, 1970).
   Useful for the detailed discussion of irreversible type II superconductors.

B. *Specialized books*

1. Saint-James, D., Sarma, G. and Thomas, E. J., *Type II Superconductivity* (Pergamon, Oxford, 1969).
2. Campbell, A., and Evetts, J. E., *Critical Currents in Superconductors* (Taylor and Francis, London, 1972).
3. Solymar, L., *Superconductive Tunnelling and Applications* (Chapman and Hall, London, 1972).

C. *Conference proceedings*

Besides the conferences on low-temperature physics, the following major conferences may be mentioned:
1. International Conference on the Science of Superconductivity, Colgate, 1963. Published in *Rev. Mod. Phys.* **36,** 1–334 (1964).
2. International Conference on the Science of Superconductivity, Stanford, 1969. Published as *Superconductivity*, ed. F. Chilton (North-Holland, Amsterdam, 1971).
3. Symposium on Applied Superconductivity, 1970. Published in *J. App. Phys.* **42,** 1–189.
4. Applied Superconductivity Conference, Annapolis, Maryland, 1972. *IEEE Publication No. 72CH0682-5-TABSC*.

Index